Coastal Change, Ocean Conservation and Resilient Communities

Marcha Johnson · Amanda Bayley
Editors

Coastal Change, Ocean Conservation and Resilient Communities

Springer

Editors
Marcha Johnson
ASLA
Brooklyn, NY
USA

Amanda Bayley
ASLA
Croton-on-Hudson, NY
USA

ISBN 978-3-319-41913-8 ISBN 978-3-319-41914-5 (eBook)
DOI 10.1007/978-3-319-41914-5

Library of Congress Control Number: 2016953302

Printed on acid-free paper

This Springer imprint is published by Springer Nature
The registered company is Springer International Publishing AG
The registered company address is: Gewerbestrasse 11, 6330 Cham, Switzerland

Foreword

The word "adaptation" is frequently mentioned in the scientific, engineering and planning literature, but is subject to many interpretations. The simple definition of adaptation as modification to meet new conditions does not indicate how this modification will be done or should be done. Actions to adapt to natural processes can accommodate or resist these processes. In the distant past, adaptation to coastal hazards occurred by abandoning infrastructure and retreating from the coast. We could compliment our ancestors on their wise decisions to remove themselves from hazards in this way, but frankly, retreat was their only option, given the power of the sea, the limited numbers of people living at the open coast and the lack of technology, economic incentive and political will to protect them. Conditions have changed greatly. Much of the coastline in many countries is densely populated and developed with buildings and transportation routes or protected by static structures. More people are moving to the coast than leaving it, and more of the coast is being converted from natural environments to human artifacts.

Recent studies have identified the need to adapt to climate change by modifying human vulnerability by flood proofing structures, evacuating during major storms, or removing buildings and infrastructure and allowing landforms and habitats to migrate landward. Other studies indicate ways to retain or restore natural values while remaining in place, using hybrid structural and non-structural solutions. Adaptation, by definition, requires some kind of human action, but many alternatives are possible. Planners and managers are often unsure about how, when and where adaptive measures should be taken. Business as usual is not an option where storm intensities are increasing, beaches are narrowing, landforms and habitats are being eliminated or truncated, natural sediment sources are dwindling, and increasing numbers of people are being placed at risk.

Only a few decades ago, scientists studying landforms and habitats considered humans as an aberration or overlay on the natural system. We must now acknowledge that humans are intrinsic to coastal systems, and our activities include both eliminating and restoring landforms and habitats. The challenge is to find ways to retain or regain natural goods and services in environments restricted by the advancing sea and encroaching human development.

Retreat from the coast is not considered the default option at present, although it may be the front-running option in the future in some locations. Economic and political priorities will be directed toward protecting human lives and infrastructure, but enhancing shore protection does not preclude adding natural resource values compatible with residential and recreational uses. Good demonstration areas are needed to document the viability of adaptation and allow people to see, use and appreciate elements of the natural environment. It will be interesting to see how much natural habitat and how many species can be protected in the attempts to reduce the risk to humans.

Karl F. Nordstrom
Rutgers University

Contents

About the Editors and Contributors

About the Editors

Marcha Johnson, ASLA—Landscape Architect, Ecological Restorationist and Adjunct Professor is a landscape architect for New York City Parks, focused on ecologically rich urban waterfronts. She has specialized in landscapes the interface of urban infrastructure with the ecosystems of estuaries, littoral zones and beaches. An adjunct professor in City College of New York's Spitzer School of Architecture since 1991, she teaches courses in the landscape architecture and sustainability programs, including Sustainable Soil and Water. She has lectured at regional and international conferences on her work addressing the incorporation of floodplain functions in urban waterfronts, adaptation to sea level rise in the context of conserving living ecosystems of coastal cities and recognizing the ecological benefits of novel, spontaneous plant communities. She is currently working on a phytoremediation demonstration project on Randall's Island in the East River, as a more sustainable alternative response to dealing with contaminated "historic urban fill" of urban waterfronts than off-site disposal in distant landfills. Marcha Johnson's written contributions to this book are her own ideas and opinions, independent of policies of either the City of New York or City College of New York.

She holds a BS in Biology from University of Illinois, a Masters of Landscape Architecture from Louisiana State University and a PhD in City and Regional Planning from University of Pennsylvania.

Amanda Bayley, Landscape Architect is a licensed landscape architect with over 10 years of experience in the field. She is focused on creating designs that are inspired and founded upon a site's role within its natural environment. Her application of ecological principles in a design helps to facilitate 'magical' moments in nature—creating spaces that have an emotional impact on the people who use them. She has worked at several landscape architecture and ecological design firms both private and public. She has a BA in Geography from Hunter College, New York, an MLA from the City College of New York, and a certificate in Sustainability Analytics from the Earth Institute at Columbia University. She recently founded Bayleywick Green, a landscape design firm dedicated to creating ecologically rich environments in the suburban landscape.

www.bayleywickgreen.com

List of Contributors

Carter Craft Hoboken, NJ, USA

Sylvia Earle Mission Blue and SEAlliance, Napa, CA, USA

John R. Gillis Rutgers University, New Brunswick, USA

Steven N. Handel Center for Urban Restoration Ecology, Rutgers University, New Brunswick, NJ, USA

Peter Hummel Anchor QEA, LLC, Seattle, WA, USA

Marcha Johnson ASLA, Brooklyn, NY, USA

Christina M.K. Kaunzinger Center for Urban Restoration Ecology, Rutgers University, New Brunswick, NJ, USA

Tricia Martin WE Design Landscape Architecture, Brooklyn, NY, USA

Andrew Moore Quennell Rothschild & Partners, New York, NY, USA

Bryan Quinn One Nature LLC, Beacon, NY, USA

Richard Roark OLIN, Philadelphia, PA, USA

Eric Rothstein eDesign Dynamics, New York, NY, USA

Anna Spooner Anchor QEA, LLC, Seattle, WA, USA

Donald Watson EarthRise, Trumbull, CT, USA

Adam W. Whelchel The Nature Conservancy, New Haven, Connecticut, USA

Chapter 1
Introduction

Marcha Johnson

A challenge facing contemporary coastal cultures is to find a way to live on shorelines that embraces, celebrates, and supports a healthy ocean. Other authors have done a very good job of exploring various architectural and engineering approaches to creating structures which can occupy flood zones so that the lower floors can be occasionally flooded with minimal damage, which we need not repeat. Instead, we address how coastal cities might adapt so that they conserve and restore aquatic ecosystems along their edges. What is good for natural environments is generally also healthy for humans: clean water, productive soil, vegetation communities that support the pollinators necessary for our food crops, healthy networks of ecosystems that allow the ocean to thrive, and also provide the ecosystem services upon which we depend.

Clean water, natural processes able to cycle nutrients and oxygen, with the fullest complement of marine organisms possible, and conserving the existing shorelines that are capable of adjusting naturally to the changing conditions which are normal to systems at the interface of earth, sea, and atmosphere, seem to be the last priority for many of the recent "resiliency" programs promoted by coastal cities. Continuing to compromise ocean-ecology is the equivalent of cutting off our own oxygen supply; ignoring the crashing populations of fish largely due to human activities and the cascading effects on other organisms including humans is as large an issue as the removal of rainforests. There is a huge disconnect between the community of scientists and environmental advocates who see the damage as well as the benefits of trying to control coastal processes in major urban deltas, and those proposing massive sea gates, seawalls, and wave breaks to reduce waves and storm surges but which would do little to prevent flooding from sea level rise [1–4].

Beyond the tragic costs and emotional upheaval of people losing their homes and community connections, responding to the human disruption caused by

M. Johnson (✉)
ASLA, Brooklyn, NY, USA
e-mail: marcha.johnson@gmail.com

M. Johnson and A. Bayley (eds.), *Coastal Change, Ocean Conservation and Resilient Communities*, DOI 10.1007/978-3-319-41914-5_1

powerful weather events is a huge drain on public resources. Federal Emergency Management Administration budgets and staffing are overwhelmed by the enormity of the tasks to rehouse people displaced by flooding, pay for repairs, and refunds on such a large scale and with such rapid repetition. More frequent and intense storms are undoubtedly on the way. Meanwhile, disregard for sustainable fishing practices, use of fishing techniques which wreck the sea bottom and inhibit the recovery of spawning, pollution, floating plastics, and warming waters have brought ocean ecosystems into great imbalance.

This is an excellent moment to reconsider the basis of former decisions based on the best current information, to draw on the wisdom of other water-edge societies, especially indigenous cultures, and engage many creative thinkers to conceive the next steps in the evolution of coastal cities—as places which conserve the rivers and oceans. Philosopher of ecology, Joanna Macy, describes the present as, "The Great Turning," a transition from a society shaped primarily by industrial growth to a society structured to be life-sustaining [5].

What practical, feasible ways are there to reverse some of the damage caused by treating the ocean as an enemy to be fought, controlled, and resisted instead of the creator of the beaches and shorelines we desire?

Concerned coastal communities are taking steps toward better coexistence with their surroundings, with varying degrees of ambition to actually support and integrate adjacent ecosystems. Some projects add a bit of artificial habitat—essentially "green window-dressing"—to otherwise traditional dikes, seawalls, levees, and revetments. Others achieve infrastructure innovations with more of the porosity, flexibility, multi-layered vegetation, and connectivity of the natural systems they displace. Many compromises affect final results—the messy consequence of working with real clients, site constraints, and conflicting agendas. Preliminary examples, small gestures, and hybrid or imperfect solutions may gradually help shift conversations about resilience toward ecosystem conservation. Building the comfort level of decision-makers and stakeholders, building a knowledge base of what works or not, and raising awareness of new approaches, is a start, setting the stage for more significant next steps.

Some possible answers can come from indigenous cultures who have lived side by side with the ocean, and in some places still do, and prehistoric cultures who left behind evidence of centuries-long adaptations to changing coastal conditions. Other answers can come from the handful of designers and engineers who have demonstrated projects which reconstruct some of the broken links, illustrate ways to adapt and have created models for innovative ways to restore some floodplain functions within urbanized waterbodies. Additional answers may come from enlightened politicians and urban planners with innovative ways to address fundamental conflicts between insuring property against damage and deliberately ignoring the hazards of building on barrier islands, beaches, wetland buffers, and floodplains. Rolling easements, for example, the "regulation or an interest in land in which a property owner's interest in preventing real estate from eroding or being submerged yields to the public or environmental interest in allowing wetlands, beaches, or access along the shore to migrate inland," is one such response to sea level rise which acknowledges the current property owners rights along with the idea of disconnecting the ocean edge from the laws for dry land [6].

Because coastal zones are different structurally and in general are more active than upland/inland areas, they require management, legal, real estate, and architectural strategies which take these differences into account. In many cities, coastal "land" is treated as the equivalent of dry land property, which is the root of many of our self-made problems. It would be more accurate and helpful to consider these ecologically transitional spaces as part of the water instead of part of the land. Locating housing, critical infrastructure, and features easily damaged by salt, water, and settling in floodplains, at elevations intermittently under water is a choice. Through the painful lessons of several recent big storms, combined with a continuing evolution of sophisticated computer models, and scientific knowledge of coastal geological activity, perhaps our culture has begun to understand that we need to "live by the rules of the Sea," which are larger and more enduring than even the strongest flood-resistant engineering [7].

Each coastal site is different with respect to its interactions with the ocean currents, waves, littoral drift, sediment transport, and erosion and accretion of adjacent land forms. Even in the same neighborhood, micro-conditions and relatively small differences in elevation or orientation to the water can significantly affect how much disturbance results from high water events.

Differences between shorelines which retain natural landforms and are covered in vegetation—for example, mangrove forests in Indonesia, or sand dunes clothed in beach grass on Fire Island, NY—experience much less impact from major weather events. Perhaps even more important from an adaptation standpoint, these kinds of coasts are able to rapidly reestablish themselves following a storm disturbance, by contrast with urbanized shores typified by seawalls, jetties, revetments, and pavements. Conserving natural vegetation that traps and anchors sediment [8, 9] can also stabilize the coastline soil in a way which allows flexibility, change and self-repair of the erosion-controlling, tangled network of roots, stems, and branches following intermittent catastrophic disturbance, a built-in hallmark of naturally vegetated coasts.

Bioengineered slopes—hybrids of living erosion control plants plus nonliving structures of wood, masonry, geotextiles, or other materials–are similarly able to regrow and repair themselves to some extent after disturbances or death of the plants in ways which concrete and steel walls can never do. Over time, such structures can become stronger as the plants get larger, as contrasted with most urban structures that become weaker due to weathering and shifting.

Conserving the natural system which enables shores to adapt and rebuild themselves includes conserving sand on and off the beach, sand-based landforms, and the sources of sand which feed beaches and dunes. Attempting to build dunes which are intended to stay put, to protect a specific piece of adjacent property, may be only as durable as the next big storm, after which they will need to be reconstructed. Regardless, in the quiet interim periods they can provide dune habitat and aesthetic benefits. These are likely to be relatively short-term features as barrier islands, beaches, and the mean high water line migrates inland. Conversely, the effort to remove overwash sand (evidence of beaches moving inland) from roads will likely be accelerated until the roads are relocated or elevated to accommodate the beach migration.

The conflict between ocean processes and modern human occupation of the shore is greatly expanded by the vast, voluntary migration of inland people to live in the world's coastal cities. Much of the tragic loss of life and expensive damage during recent storms is due to the deliberate construction of housing in low lying areas within the past century and especially the past 50 years. Further, efforts to not only rebuild "permanent" structures in the same known hazard zones, and many new buildings on shorelines not only continues, but continues at an accelerated pace in coastal cities around the world. Densely concentrating people in areas highly susceptible to extreme events of wind, waves, storm surge, flooding, and coastal sediment movement—erosion and accretion—can no longer be considered accidental, unintended or the result of a lack of knowledge of coastal geology. Research shows that losses from natural and related technological hazards are often the consequence of narrow or shortsighted development patterns and an attitude toward the natural environment [7] as a nuisance to be managed and controlled. The economic incentive to build high rise hotels on the beach (for example in Atlantic City and Miami), reconstruct low income housing on barrier islands (The Rockaways in New York City), as well as the understandable but shortsighted emotional incentive to "come back stronger than ever" perceived by many as gritty optimism, instead of obstinacy in the face of obvious error, calls for some new models of science-guided, sustainable, ocean-ecology-conscious approaches to living near the shore.

Language of Fear and Culture of Emergency Response

The use of language to influence behavior has been used time and again. The media in particular is adept at capturing the public's attention by employing terms of fear and violence to heighten emotional responses to news stories, and the language used to describe weather events is particularly angled to get our attention. Using exaggerated terms to get us to watch the news is a fairly minor issue. However, it becomes more serious when such emotional triggers as the language of fear, disaster, and emergency affect or even determine decision making, and this seems to be the case in many choices regarding adaptation to coastal change especially sea level rise. As the Earth is experiencing major changes in climate, sea level, species creation and extinction, we would greatly benefit from rational, thoughtful, and science-based decisions about how to respond in general and how to behave on coastlines in particular. The mindset which prevails during fear and emergencies—

panicky reactions, the tendency for politicians to want to show that they are responding by throwing a lot of money at the problem, the search for simple one-size-fits-all solutions, extreme stress, limited time to think, and the sense of needing to act fast—is getting in the way of more creative and evolving approaches to figuring out how to not only survive within the changing environment but to live and thrive near the ocean without destroying it.

By describing typical, natural events such as storms, rain, flooding, snowfall, waves and the climate using the language of fear to describe them, such as: "threatening, dangerous," or violent terms such as "brutal winter storm" or "the uncontrolled fury" of a river, and exaggerated to seem to have bigger impact than normal such as: "massive" flooding (describing common winter weather and typical, flooding within a known floodplain) as "extreme" events can have the cumulative effect of disconnecting people from a rational, healthy respect for the power of nature. This language instills the perception that: (1) nature is an enemy, at least from time to time, and needs to be controlled or overpowered; (2) protection of flood-prone coastal property is something that the government is supposed to do for its citizens; and (3) one is entitled to remain living in a flood zone as sea level rises, as if it were upland, because one has purchased and owns the property.

A related language-induced perception is the result of federal funding being made available during times of emergency. There is an incentive by states and municipalities to declare "states of emergencies" for some fairly predictable, repeating but temporary situations which, yes, can be dangerous especially if people are underprepared but really only "emergencies" of our own making. These include events such as a few days of very heavy rain or a snowfall which exceeds the capacity of the local snow clearing operations, or flooding in developed urbanized floodplains. In July, 2015 the Pentagon issued a report that climate change is a security risk and impacts national security interests around the world, using language similar to that describing terrorism. "The report finds that climate change is a security risk, Pentagon officials said, because it degrades living conditions, human security and the ability of governments to meet the basic needs of their populations" [10].

The cost to ocean ecosystems of human consumption and management of resources has been huge, and one result of our interventions has been a trend toward beach-less coastal cities. One model of urban shorelines, exemplified by Nice, France, is the festive paved esplanade along a seawall paralleling the shore, where previously there were wide, gentle sandy beaches but now are steep pebble beaches moved back into place by bulldozers after waves and currents have shifted them out of the intended alignment. In other places, such as Waikiki, Hawaii there is very little beach left seaward of the seawall. The beach-side hotels are still there, just without the beach. In such examples, the former transitional, flexible, and responsive soft shores have been unable to migrate inland with natural processes. Because the seawalls have minimal habitat value and almost no capacity to accommodate floods, and store water, the living and nonliving ecosystem services of a beach have been displaced, in addition to the beaches that attracted tourism. There are better ways to live beside the ocean. In the following essays, we explore some.

References

1. Marshall C (2014) Massive seawall may be needed to keep New York city dry. Climatewire May 5, 2014. http://www.scientificamerican.com/article/massive-seawall-may-be-needed-to-keep-new-york-city-dry/. Accessed 24 April 2016
2. Maeslantkering https://en.wikipedia.org/wiki/Maeslantkering. Accessed 24 April 2016
3. Aboriginal Peoples Television Network ATPN (2014) Venice tests massive movable flood barrier. http://www.telegraph.co.uk/travel/destinations/europe/italy/venice/10376184/Venice-tests-massive-movable-flood-barrier.html. Accessed 24 April 2016
4. Johnson M (2012) An evolving coastal city. www.Coastalcare.org
5. National Public Radio Program (2012) On being, interviewing writer Joanna Macy. http://www.onbeing.org/program/wild-love-world/transcript/4905. Accessed 24 April 2016
6. Beachapedia (2014) Rolling easements. www.beachapedia.org/Rolling_Easements. Accessed 18 April 2016
7. Mileti DS, Gailus JL (2005) Sustainable development and hazards mitigation in the United States: disasters by design revisited. In: Mitigation and adaptation strategies for global change, vol 10. Issue 3, pp 491–504. http://link.springer.com/article/10.1007%2Fs11027-005-0057-4#/page-1. Accessed 18 April 2016
8. Bush DM, Pilkey OH, Neal WJ (1996) Living by the rules of the sea. Duke U. Press, Durham and London, pp 2–16
9. Ibid, 17
10. U.S. Department of Defense (2015) DoD releases report on security implications of climate change. http://www.defense.gov/News-Article-View/Article/612710. Accessed 18 April 2016

Chapter 2
Traditional Cultures Editorial: Mobility and Innovation in Traditional Coastal Cultures

John R. Gillis

It may come as a surprise to know that before the modern era, the past weighed much less heavily on the present than it does today. Traditionalism, a mindset that pays obeisance to the past, is in fact more characteristic of our own times than earlier coastal cultures. The kind of nostalgia associated with modernity prevents us from confronting coastal problems with the pragmatism characteristic of our ancestors [1]. What is striking is just how flexible and innovative they were.

The traditional coastal cultures referred to here existed from the end of the Ice Age to the middle of the nineteenth century, when the current sea rise began. By 4000 BCE, coasts stabilized and assumed roughly their current shape around the world. For the next six millennia there was relative constancy, though this was frequently punctuated by storms, floods, and tsunamis. But since 1850 a second epochal inundation has been detected, and, coupled with increased temperature levels, has brought further perturbations that recall earlier epochs of sea rise [2].

Before the nineteenth century historical consciousness did not run very deep, and the future did not loom large. People's lives were narrowly focused both spatially and temporally. History was inseparable from communal memory, which

J.R. Gillis (✉)
Rutgers University, New Brunswick, USA
e-mail: gottgillis@cs.com

M. Johnson and A. Bayley (eds.), *Coastal Change, Ocean Conservation and Resilient Communities*, DOI 10.1007/978-3-319-41914-5_2

was highly selective. It was passed from one generation to the next, was always open to innovation, responsive to changes in the experience of the environment itself. This was particularly true of highly mobile hunter-gatherer peoples, but also of settled agrarians who arrived on the human scene about 9500 BCE. In time, agricultural surpluses would contribute to the creation of kingdoms, cities, and priesthoods. With the arrival of mass literacy in the nineteenth century, local experience would give way to expertise. In our time, people are much less free to innovate and knowledge systems have become more rigid, less responsive to immediate, local circumstances.

Recent research has shown that the original *Homo sapiens* were coastal dwellers, originating on the shores of South Africa roughly 164,000 years ago in response to the drying out of the continent's interior [3]. At the shore, humans discovered a far richer environment, conducive to brain development, tool and language invention, and new forms of kinship focused around a common hearth. It is not unfair to say that the shore was the original home of what we regard as our own species [4]. From that point onwards, human evolution took shape along the shores of Africa, Asia, Australia, Europe, and the New World, following what has been called the great "kelp highway" alongshore rather than toward the interiors [5].

While coastal people ultimately colonized the interiors of Europe, Asia, and the Americas, I am reluctant to call them "settlers." That was not the original intention of their movements. Their migrations were alongshore, not inland. Where land and water met they encountered what we call an *ecotone*, a hybrid composed of two different ecosystems of exceptional diversity and richness. This special place generated a mixed population of fishers, gatherers, and farmers, a unique coastal people that have existed almost to the present day. They were distinct from landed populations, but were not exclusively maritime. The tendency of both historians and anthropologists to focus almost exclusively on either peasants or mariners has caused coastals to be largely neglected. Geographers have also ignored shores in favor of interior lands, but we are now coming to see them as not just the edge of land or sea but a third space worthy of close examination. Recent studies of human evolution now insist that *Homo sapiens* have not followed a single line of development but several. To understand the world as we know it, we must understand the diversity of our species as this developed in response to various ecological niches.

Ultimately a majority of *Homo sapiens* would turn to agriculture, what the anthropologist Yuval Noah Harari has called "history's greatest fraud." [6]. According to him, agriculture forfeited the kind of Stone Age affluence enjoyed by coastal hunter-gatherers and pastoralists. Peasants toiled more hours, were exposed to higher levels of disease and mortality, and enjoyed less freedom. They were also more class and gender divided. Coastal peoples would remain a minority into the modern era, but their geographical marginalization should not be mistaken for inferiority or backwardness.

Recovering this amphibious history is not easy, but it is important to note what makes coastal people unique. Like all people who occupied something in-between, they are both vulnerable and privileged in important ways. They mastered a geography that is best likened to a frontier, without clear boundaries. Shores are themselves in constant motion and it is this which sets the coast apart from the interior.

Coastal people have always been freer than inlanders, less subject of kings and priesthoods. They have a longstanding and well-deserved reputation for lawless and irreligious behavior. Until the nineteenth century, coasts and their denizens were perceived as dangerous, even savage. Like nomads and frontier peoples everywhere, they were threats to order organized around fixed boundaries, private property, and static settlements. Like the sea nomads and pirates with whom they were often confused, coastal people were set apart by their very high rates of mobility. They were the last humans to be domesticated and they retained those precious qualities of flexibility and resiliency that are in such short supply today.

Fundamental among humanity's original adaptive capacities was mobility itself. The original colonization of African shores was the product of the massive drying out of the continent roughly 160,000 years ago. Movement to the sea opened up a range of food resources that would ensure the future evolution of *Homo sapiens*, whose larger brains and enhanced cognitive abilities made possible migration out of Africa, eventually populating both the old and new worlds. Before the end of the Ice Age, when the seas were shallower, this was accomplished largely by moving along coasts. But when deep water wayfaring skills were developed even the most isolated island peoples were able to avoid environmental disasters by island hopping, transporting not only themselves but their animals and plants across vast stretches of the Pacific. Hunter-gatherers on the coasts of Alaska and elsewhere were equally skillful at moving alongshore and inland to avoid natural disasters. Today, their access to similar strategies is inhibited by property laws which have prescribed the ancient rights of the wayfarer as well as those of the hunter-gatherer.

As we usually tell the story of human progress, it begins with foraging, proceeds to agrarian settlement and then, recently, to urban development. We think we have domesticated nature, beginning with crops and animals, but in reality we have domesticated ourselves, turning indoors, barricading ourselves within the walls of the *domus*, closing our doors to both nature and the outside world. Thinking we have achieved security, we have gained only its illusion. And in the course of domesticating ourselves, we have reduced our resiliency by a considerable degree.

As Rachel Carson famously put it, "the edge of the sea remains an elusive and indefinable boundary." [7]. There are not only the temporal transformations wrought by daily, monthly, and seasonal tides, but where shores are composed of sand or mud, another fluid element. Sand moves alongshore, stopping to rest but rarely standing still for very long. Coasts were the last places to be mapped in detail; and the "coastline" was not identified and objectified until the late eighteenth century. Until then, certain landmarks, headlands and bays, were known to mariners. Until the modern era, the greater part of seafaring was shallow rather than deep water, along coasts rather than between them.

The boats employed in the coasting trade were shallow-bottomed, built for river as well as ocean traffic. They were easily drawn up on shore; and temporary landings far outnumbered permanent harbors until the nineteenth century. Ports with deep harbors were located largely inland, upriver, where they were more easily defended against storms and pirates. We imagine shores to have been lined with fishing villages, but, in fact, these became prominent in Europe only in the seventeenth and eighteenth century. Until then, peasants came down to the shore when the fish were running, keeping there only storage sheds for their gear. They camped rather than settled on the shore, deliberately choosing flimsy, floatable constructions that could be easily abandoned or moved out of the way of storms, tides, and tsunamis.

In contemporary Maine, the shore cottage is still called a "camp," a clue to the transient, even nomadic nature of earlier coastal generations. Brian Fagan writes: "Generational memory and cherished oral traditions would have underlain a way of life defined by movement, not environmental change." [8]. Like pastoral or nomadic people to whom they should be compared, the fishers/farmers (often one and the same) of coastal regions moved freely across wide areas their lives focused on a few points of safe refuge, but entirely oblivious to the kinds of boundaries which had come to define the lives of agrarian peoples. Their seafaring was akin to the wayfaring of the aboriginal peoples of North America and Australia, reliant, like the birds and animals whose tracks they followed, on natural landmarks rather than instruments or maps [9]. They had their marine version of aboriginal songlines, providing them with a sense of viable routes, unburdened by the notion of roots that would inhibit the movements of later generations of people tied to land rather than water.

Leaving behind few written records and even fewer monuments, coastal history has remained obscure in comparison to its better documented interior. Coastal people had a much more fluid sense of time and space, capable of moving out of the way of danger presented to them by the sea. We can trace their flexibility back to skills long practiced by hunter-gatherers. Yet, historians and anthropologists continued to insist that the Neolithic Agricultural Revolution was humanity's great leap forward, a myth reinforced by the Biblical story which would have us believe that everything begins with the Garden of Eden. The truth is that the simultaneous domestication of humans, plants, and animals exposed our species to pollutants and diseases which peoples on the move rarely encountered. Until the emergence of modern medicine and sanitation, movement had been the only way to reduce mortality. In this case, low mortality pared with low fertility contributed to a way of life that was in many ways preferable to that inland, despite the obvious dangers posed by the ocean itself.

For much of the agrarian era, coastal dwellers were relatively few in number and sufficiently mobile to avoid mass disasters, but all this has changed with modern population rise in the nineteenth century and now with the onset of further sea rise. Originally unnoticed by observers, sea rise coincided with a historically unprecedented surge to the sea by inland populations around the world. Colonized initially and lightly from sea, coasts are now overrun by inlanders who want to live *on* the sea, but have not the slightest idea of how to live *with* it. Industrial fishing and the rise of container shipping has reduced the numbers of working waterfronts of all kinds, displacing them from shores that have been taken over for recreational purposes. The shore has become thoroughly domesticated and urbanized, and the once mobile camp became a "teardown", allowing the shore property to rocket in value. To protect such assets, the shore has become a place of objectified, engineered boundaries, bounded by massive sea walls which are generally agreed to be the cause rather than the solution to erosion and storm damage [10].

The disasters we have experienced in the last two decades are not natural, but human in origins. What was once a fluid frontier, a broad and flexible margin bridging land and sea, coasts have become increasingly narrow and intransigent edges, full of unforeseen dangers. Vernacular knowledge has been cast aside in favor of engineering expertise based standards developed for terrestrial rather than aqueous environments. In the era of globalization and the nation state, local people have lost control of the shore, for what was once a frontier, a realm of freedom of movement, is now a closely guarded coast. For the people who have experienced Hurricane Katrina or Sandy, their paradise has become a purgatory [11]. The values of the landlubber have triumphed everywhere, although, in some of the least developed areas of the world, like Somalia, it still possible for indigenous people to assert themselves through piracy. England's coasts are sprinkled with names like "Smugglers Cove," which recall a different kind of coastal culture, but there will be no return to the days when the shore was the castaway's refuge and the migrant's last resort.

Tourism has invested the shore with nostalgia, but we must hold up the true history of our coasts as a mirror to current conditions if we are to get a viable perspective on our contemporary dilemmas. We need to recognize that we are essentially the same human species that first came down to the African shores; we need to remember that sea rise has happened before without resulting in total catastrophe or extinction. We need to appreciate the examples of adaptation that coastal peoples of every place and time over the last fifteen millennia can offer. There is nothing worse than to condescend to our ancestors, especially when the modern era has erased so many of the adaptive skills that they bequeathed to us. Having replaced so many traditional navigational and fishing skills with mechanical devices and electronic aids, it is we who have lost touch with the ways of nature as well as our natural selves.

There is so much to be learned from the coastal past, from the mound builders of the northern Europe as well as the stilt houses of the Pacific. There is the history of the so-called Sea Peoples of South Asia, many of them refugees from mainlands wrecked by war or famine. While Sea Peoples almost never spent their entire lives on the water, they can show us how to live between land and sea, drawing on the resources of both without devastating coastal environments. We need also to take into account the history of waterlands, estuaries, and wetlands, which for millennia were among the most productive places on earth until marginalized by the drainage and dyking projects of recent times. Here we can look to the Netherlands, the European country with the longest experience with inundation, where the learning curve is sharply up. Massive engineering projects have been supplemented by experiments with floating buildings, including schools, prisons, and factories. Restoring wetlands and creating soft edges between land and sea is another strategy that draws on past experience. It seems it is not too late to enlist ancestral wisdom in our inventory of coastal solutions [12].

In this new era of sea rise, we have the opportunity to return, as the first *Homo sapiens* did, to the shore and to its past in order to use the lessons it can teach us. The coast can again be a frontier that we move with and adapt to. Embracing fluidity over fixity requires rethinking many of our most deeply held spatial and temporal myths, but humankind has shown its genius for adaptation before and can do it again if it can summon up the political will.

References

1. Gillis JR (2010) Nostalgia by the sea. Island Journal 26:27–31
2. A useful summary of sea levels is provided by Brian Fagan (2013) The attacking ocean: The past, present, and future of rising sea levels. Bloomsbury Press, New York
3. Marean CW et al (2007) Early human occupants use of marine resources and pigment in South Africa during the middle Pleistocene. Nature 449:905–908
4. Sauer CR (1963) "Seashore—Primitive home of man", in Land and Life: A selection from the writings of Carl Ortwin Sauer. University of Califronia Press, Berkeley, p 309
5. Gillis JR (2012) The human shore: seacoasts in history. University of Chicago Press, Chicago (chapter I)
6. Harari YN (2015) Sapiens: a brief history of Humankind, Harper, New York (chapter 5)
7. Carson R (1955) The edge of the sea. Houghton Mifflin, Boston, p 1
8. Brian Fagan (2013) The attacking ocean: the past, present, and future of rising sea levels. Bloomsbury Press, New York, 37
9. Ingold T (2007) Lines: a brief history. Routledge, London, pp 75–77
10. Pilkey O, Young R (2009) The rising sea. Island Press, Washington DC
11. Monmonier M (2008) Coastslines: How Mapmakers frame the world and chart environmental change, University of Chicago Press, Chicago (Gillis JR (2015) The shores around us, Createspace, Charleston, 2015, chapter III)
12. Gillis (2006) The Human Shore, chapter 6. On sea peoples and South Asian littorals, see Michael Pearson, Littoral society; the concept and the problems. J World His 17.4:353–73. (On the long history of innovation in the region now known as the Netherlands, see Robert J. Hoeksema, Designed for Dry Feet: Flood Protection and Land Reclamation in the Netherlands, Reston, Virginia: American Society of Civil Engineers, 2006, chapters 1–3)

Author Biography

John Gillis—Historian, Author, and Professor John Gillis is a Rutgers University Professor Emeritus of History who resides for most of the year in Berkeley, California. In the summer, he can be found on Great Gott Island, off Acadia National Park in Maine. He has taught at Rutgers, Stanford, Princeton, and the University of California at Berkeley, and is a Life Member of Clare Hall, Cambridge. Mr. Gillis has been a Visiting Fellow at St. Antony's College, Oxford, and was a fellow at the Woodrow Wilson International Center for Scholars, the Center for Advanced Study in the Behavioral Sciences at Stanford, and the Swedish Collegium for Advanced Studies in the Social Sciences at Uppsala University. He is also a Fulbright Senior Scholar. John Gillis is the author of numerous books and other publications including *The Shores Around Us*, 2015; *The Human Shore: Seacoasts in History* (2012) and *Islands of the Mind: How the Human Imagination Created the Atlantic World* (2004).

Dr. Gillis holds a B.A. from *Amherst College* and a Ph.D. from *Stanford University.*

https://en.wikipedia.org/wiki/John_Gillis_(historian).

http://johnrgillis.com/Biography.htm.

Chapter 3
Designing the Edge Editorial

Carter Craft

How does one insert consideration of ecological values into a development-centric waterfront?

The post-industrial landscape of New York City offered thousands of acres of land and a hundred miles of waterfront for new uses. Over the latter half of the twentieth century, as many factories, large and small, moved to other parts of the northeast, the US or around the globe, large stretches of New York City's waterfront (e.g. the Hudson River waterfront) became available for new uses. In the first

C. Craft (✉)
Outside New York, LLC, 608 Garden Street, 1st Fl., Hoboken, NJ 07030, USA
e-mail: carter@outsidenewyork.net

M. Johnson and A. Bayley (eds.), *Coastal Change, Ocean Conservation and Resilient Communities*, DOI 10.1007/978-3-319-41914-5_3

generation of new parks and public spaces to emerge, however, the water was generally seen as a visual amenity. Design that interacted with the water or fostered aquatic habitat was very rare. It was not a design priority. As awareness of larger environmental issues arose concerning the Hudson River and concern about climate issues around the globe began to grow, the second generation of waterfront projects that began in the early twenty-first century looked to create many more benefits for recreation while improving the environment. This chapter reviews the shift in recent waterfront design thinking as more stakeholders began to understand and appreciate the biological complexity that that waterways surrounding New York City represent.

Like many other great cities around the world, New York City grew in part because of its location near the mouth of the Hudson River in Upper New York Bay. To the southwest of the city, the Raritan River, which is the second largest river feeding the great estuary, empties into the Lower Bay. The network of more than a dozen large and small rivers that feed the Hudson-Raritan estuary make this area very biologically diverse, with numerous different species of fish routinely migrating, feeding, and nesting here. From the earliest times of human settlement, these waterways have been seen as edges between areas and between places. For example, in New York, the Harlem River creates the boundary between the Bronx and Manhattan, while the Hudson River is the boundary between New Jersey and New York.

About Edges

No matter the location, edges are always physically complicated conditions. Start with what you are wearing right now. The cuff is often the most stressed and first point of failure on your favorite pants or shirt. At the edge of the carpet is where things start to fray. Outside the edge of the sidewalk is a danger zone with traffic zooming by, at the curb is where the edge of the sidewalk and the edge of the street both start to fail. On an urban design scale, the elevated highway or the railroad track is considered the edge of the neighborhood. Sociologists talk about the "fringe" of society, referring to populations at risk or on the edge of society. However, in terms of local climate and ecology, the land's edge is where the heat island effect disappears. Estuaries at the edge of the oceans are some of the most biologically productive areas on the planet.

In terms of human sensibility, edges are also often a gray zone of responsibility and often a lower priority. I have been a part of many meetings where the responsibility for street trees was debated and passed back and forth like a buck between the agency responsible for asphalt and the agency focused on greening. On a real estate level, property owners generally pay much more attention to the building assets than they do the sidewalks in front. At the water's edge the thinking is often the same.

Environmental Policy Seeks to Transcend the Edges

With the development of the industrial economy in nineteenth century America, many barriers grew up at the water's edge. Over time, long walls of factories, yards, and transportation hubs prevented access for anyone but the workers. Rivers, creeks, canals, and bays were used as open sewers, for residents as well as industries. This condition went on for hundreds of years.

Eventually, the Clean Water Act of 1972 forced American industry to begin cleaning up its act and particularly its waterways. Every pipe and every discharge became regulated. Every municipality larger than 50,000 people was required to construct sewage treatment plants to better process the human waste. In the polluted, essentially poisoned harbors such as New York Harbor, the industrial tide was turning. The public appreciation for the waterfront—not just at the coast but also in the urban core—was becoming a larger goal all around the nation, evidenced by the passing of the federal Coastal Zone Management Act in 1972.

In addition to the regulation and reduction of industrial discharges, the construction of sewage treatment plants throughout New York City and northern New Jersey in the late twentieth century helped marine life recover. There were and are many legacy contaminants in the silt and sediments along the bottom of the harbor estuary. Regardless, the tremendous reduction in pathogenic bacteria and the huge increases in dissolved oxygen, combined with the enormous influx of salt water and biodiversity from the incoming tide, have led to a dramatic rise in the volume of aquatic life and the productivity of the estuary for greater ocean life. Now that the water quality is improving, we begin to think, what else can be done for the marine environment to sustain this recovery?

A huge investment has been made in building sewage treatment plants (estimated at close to $40B in NYC alone). However, the habitat quality of our estuary has been almost irreversibly damaged over the past 200 years by the destruction, filling, and hardening of shorelines. There is actually very little intertidal zone left. This is unfortunate because just as estuaries are made biologically productive by the mixing of fresh and salt water, so are intertidal zones made more productive by the alternating cycles of inundation by water and exposure to air. When made by nature, these zones are gently sloping shorelines that can extend dozens and even hundreds and hundreds of feet inland and upland. In a semi-enclosed area like Jamaica Bay or the NJ Meadowlands, the intertidal zones can extend thousands of feet.

Along an urban river or canal, however, the edge condition has been hardened and made into a vertical bulkhead, or sometimes, a sloping riprap of boulders. Neither of these conditions enable the benthic or bottom zone to be a healthy part of the food chain. For example, horseshoe crabs—which have survived this planet for millions of years, today rarely if ever come ashore to lay eggs. Crabs now have few protected places to live. Wading birds have limited shallow water habitat for feeding. Small minnows (like killifish) have no shallow areas to escape larger predators. Thus, with the decline of members at the bottom end of the food chain, there is less food for the animals higher up. Biodiversity and marine life have declined.

Today, most waterfront property owners in the urban core do not need the physical link to the water that a pier or strong bulkhead provided 50 or 100 years ago. Waterfront land uses are more often land-oriented, with vehicles such as the car or truck the central component for mobility and access. For these activities the water's edge is not a basic structural element of their property holdings. The edge is of lower priority because it is not critical for day-to-day operations.

At the same time, without a bulkhead or engineered edge to keep the water at bay, the options for economic use of the land might be narrowed. Floodwater can cause damage such as undermining building foundations and ground floor activities. While keeping the water out is obviously necessary, providing or creating intertidal habitat along waterfront property is generally a non-starter; this reduces the potential footprint of a building, and thus the land's economic potential.

Given that many waterfront sites in the urban core have historic contamination at some level, there is pressure on developers to generate more money with the new land use in order to cover site remediation costs. Shrinking the footprint of a building on a waterfront lot makes it that much harder to generate the funds to cover these costs. As a result, we must work harder to achieve ecological productivity along these watery edges. In New York City, there are not many areas where we have both the upland physical space and the framework that allows marine life and biodiversity to become public priorities that are drivers in the design process.

A Moment in Time, and Some Limitations of Urbanizing the Floodplain

In 2001, Manhattan was in the early stage of a waterfront renaissance. The new neighborhood of Battery Park City had been built over the previous 20 years, with the waterfront being the first zone completely built out. Residents and visitors here enjoyed a new experience for waterfront access. Parallel pathways served active and passive users. The two paths were gently grade separated and divided by a row of healthy trees. Sun shone down along the water's edge for cyclists and joggers, while walking and sitting areas were located in the shade upland of the waterfront promenade.

The grade separation was not just between the two pathways, in fact the whole area of Battery Park City had been built above the 100 year flood elevation, meaning that the connection to the Hudson River was over a chest high railing, with the River on the other side 6–12 feet below the level of the promenade. When the attack of 9/11 happened, this vertical disconnect—between the elevation of the promenade and the elevation of the river—became a huge obstacle. Emergency responders arriving aboard boats had a sheer wall to scale, with few openings in the railing to access the promenade. Residents and workers who were evacuating had a tremendous and dangerous "leap of faith" to make, over railings and down to the decks of boats there to whisk them to safety. Suddenly this celebrated promenade was a huge linear obstacle, posing a large threat to the recovery and evacuation. Here in this polished new neighborhood, the events of 9/11 made it clear New York needed a new model for the water's edge.

Meanwhile, below the surface of the water, life was returning. The ecological trend throughout this period was tremendous recovery. The City had developed the North River Wastewater Treatment plant on the west side of Manhattan, which opened in 1986. For the first time in history, nearly all of Manhattan's sanitary sewage was processed before the effluent was released into the water. Field data collected by the City as well as observations at local environmental centers such as the River Project in Tribeca showed that fish populations were continuing to recover. As John Waldman famously wrote, there were "heartbeats in the muck" [1]. The marine life was growing stronger. And multiplying.

The same vertical condition that limited the maritime response on 9/11 also posed a barrier to greater public understanding of the river. At Pier 26, the River Project had managed to create a grassroots environmental center over 15 years despite a number of obstacles. The pier platform was first an impoundment lot made available by the State Department of Transportation. The facility, high above the river, suffered the same vertical disconnect from the water that almost every neighborhood suffered.

In the face of these contrasting conditions—an estuary in rapid recovery and vertical waterfront edges designed to prevent access, not enable it—a new awareness was emerging. Water quality was becoming predictably safe enough that human use and enjoyment was again possible and a new generation of infrastructure needed to be developed; both to facilitate human contact with the water and, hopefully, at the same time to facilitate the re-growth of marine and intertidal habitat.

Goals Were Changing, Design Approaches Were not

As redevelopment of the post-industrial waterfront continued in places like Hudson River Park in Manhattan, Queens West in Long Island City, and DUMBO and the Eastern District Terminal areas of Brooklyn, there was not much clamor for ecological productivity or great concern about sea level rise. The primary public desire to be along the waterfront was visual and closer physical access. People wanted to get closer to the water. Regardless, some designers succeeded in making quiet but significant strides to restore ecological habitat in tandem with new waterfront development.

It was only after the first generation of these new public spaces came on line that people began to see that the new infrastructure was as limiting for human activity as the old infrastructure. Yes, the chain link fences had been lowered and replaced by decorative railings, often topped in green heart or other rain forest timber. The parking lots and once-empty staging yards had been replaced by grass and plantings, often sustained by sprinkler systems. The pendulum had swung far back from the industrial era. Gray and rusty replaced by shiny and green. Yet as these new spaces opened, there was a feeling that city dwellers needed a type of access that was more engaging with the water and better integrated with the environment. The grassroots groups like the Downtown Boathouse and Gowanus Dredgers were facilitating this access, but the City itself could not replicate it.

While some people were critical of the new waterfront design du jour at this time, the design conversation and specifically the design palette that was available for urban waterfronts was quite limited. Bulkheading was expensive, but it was a generally favored because the usable space upland was kept intact and the lifespan of the bulkhead would be long. Riprap embankments could be installed in some places, but they were less in favor because creating a sloping shoreline generally meant a wedge of space upland would have to be given up.

Under zoning laws, the size of a building's maximum allowable footprint is directly related to the size of the buildable lot. Therefore giving up 10–15 feet along one side of a parcel to accommodate a sloped embankment down to the water could impact the economics of the overall project. It is also significant that under the Clean Water Act today, it is almost impossible to fill in the waterfront or change the configuration of the shoreline. This is especially true in New York. Thus, people who own waterfront land want to keep the parcel intact—they know that the larger the parcel, the larger the future value as set by a per square foot basis. Even public agencies who control waterfront land do not want to give up space. Larger blocks and larger building sites generally allow for a wider range of development opportunities. For local government this means larger potential tax revenues as well as a wider variety of potential urban activities.

Looking Back and Looking Forward Along the Waterfront Edge

Looking back, who envisioned that the industrial era would have vacated the city as quickly and completely as it has? Looking ahead, who knows what types of needs the city might have for mobility, infrastructure or other uses in the future? Therefore, it is a practical public policy to keep the waterfront land holdings as intact as possible. Clearly, there were land use and economic pressures when it came to revitalizing the waterfront. But finding the design inspirations or design precedents that could be applied along the water's edge was not clear at all.

As a society we do not like to take chances or experiment with infrastructure. Developing a new potential model for the water's edge in New York was the result of a confluence of many factors. Some of these factors were internal aspects brought to the discussion by leaders of the community as well as the agency with responsibility for maintaining the space. These factors included:

- Recognition within the lead agency that a new model for edge design was worth pursuing;
- Recognition by the project team that a new approach could inform other projects;
- Experience within the lead agency with alternative and imaginative design; and
- Availability of a site where the community was open to new ideas.

Other factors important in enabling alternative approaches for waterfront design were external, not specific to the community or to a particular place. These factors included:

- Standing of stakeholder groups to help the city develop an innovative approach in this particular location;
- Awareness among local residents that new models could be preferable and more productive than standard approaches; and
- Awareness among the wider public that the new era of waterfront public spaces needed to be cognizant of global environmental climate change.

Thus, many values were shared by both core stakeholders (i.e., those with control and significant influence over the site) and outside stakeholders (i.e., those who could lend various types of support to the effort). The entire contingent of stakeholders to re-imagine New York City's waterfront, which included four public agencies and nearly a dozen civic and non-profit organizations, ultimately developed and executed a thorough public engagement process to create a new model for waterfront edges. The project, co-funded by the NYC Department of Parks and Recreation and New York State Department of State, Division of Coastal Resources, was called *Designing the Edge*. I will not go into detail on the guidelines and content of this project—it can be accessed here: https://www.nycgovparks.org/sub_opportunities/business_ops/pdf/designing_the_edge_4-7-2010.pdf.

Instead, this chapter's review of a century of waterfront planning is intended to reveal how the next century of waterfront design may coalesce. Fortunately, a city as large and vibrant as New York functions as a living lab in many ways. The arts, media, education and many fields of design find New York a welcoming canvas for new ideas. New education models are constantly developed and tested in its more than 1000 schools and dozens of colleges and universities. New aspects of art and media are constantly being generated and consumed. The built environment can undergo a significant amount of experimentation. Waterfront edges, critical infrastructure elements that are now being designed to protect the land will increasingly in the future be designed to attract and sustain more diverse life.

Reference

1. Waldman J (2012) Heartbeats in the Muck, the history, sea life and environment of New York Harbor. Fordham U. Press, New York (Revised edition)

Author Biography

Carter Craft—Urban Planner, Waterfront Advocate with close to 20 years of experience specializing in waterfront and transportation issues. Through his planning and project management firm, Outside New York, Carter has served as a consultant to a wide range of public and private organizations. As an outdoor enthusiast, he brings a deep understanding of the complex issues inherent in planning within the ecological and political systems entangled in the waterways of the northeast. His constructive working relationships with many waterfront stakeholders and decision-makers help to develop potential solutions that are both practical and progressive. Over the last 15 years Carter has been involved in creating and growing more than two dozen water-related non-profit organizations including the Waterfront Alliance, Gowanus Dredgers Canoe Club and the Hoboken Cove Community Boathouse. His work blends innovative waterfront designs and programs with a high degree of public participation.

Mr. Craft holds a BA in Economics from UNC-Chapel Hill and a Masters of Urban Planning from NYU's Wagner School of Public Service. He is a USCG licensed Captain (50 ton) and a tour guide for organizations including South Street Seaport and the Municipal Art Society of New York.

https://outsidenewyork.wordpress.com/about-carter-craft/.

Chapter 4
Literature Review: Principles and Practices of Coastal Adaptation in the Era of Climate Change

Donald Watson

D. Watson (✉)
EarthRise, 54 Larkspur Drive, Trumbull, CT 06611, USA
e-mail: EarthRise001@SBCglobal.NET

M. Johnson and A. Bayley (eds.), *Coastal Change, Ocean Conservation and Resilient Communities*, DOI 10.1007/978-3-319-41914-5_4

Introduction

Coastal land and waterways comprise a system of co-dependent ecosystems, salt and freshwater zones that include richly diverse zones for food, nursery, and shelter for land and marine life (Fig. 4.1).

Coastal zones are vulnerable to climate change impacts of storms, rising waters, warming waters, pollution, and acidification. Sea level rise adds complexity to planning challenges, with uncertainty entirely dependent upon global climate change and coastal development. Coastal adaptation is an emerging discipline to plan for climate-related threats to coastal zones. The term "adaptation" was introduced in the mid-1990s to describe a set of strategies to address global warming and related risks, including extreme storms, flooding, and sea level rise [1].

This chapter reviews new approaches to coastal adaptation reported in recent literature. The referenced reports outline criteria, tools, and applications. Together, they represent innovative planning principles of the emerging discipline of coastal adaptation. In particular, the literature review is presented under nine working principles, highlighting unique approaches to the developing discipline of coastal adaptation:

(1) *Plan for uncertainty*
(2) *Plan with a range of options*
(3) *Address social, economic, and environmental benefits*
(4) *Integrate multidisciplinary knowledge and outcomes*
(5) *Combine green and blue agendas*
(6) *Plan for impacts from coastal water to upland communities*

Fig. 4.1 Coastal ecosystems defined by natural geology and soils, bathymetry, vegetation, and hydrology [11]

(7) *Plan for living shoreline ecological zones*
(8) *Evaluate site infrastructure and building suitability*
(9) *Integrate hydrology, flooding, and stormwater infrastructure.*

(1) *Plan for uncertainty*
The issue of uncertainty can present a significant barrier to establishing plans of action. Underestimating sea level rise will result in failing measures that increase, rather than decrease, risks along with the loss of investments thus expended. Building overcapacity—sufficient to address the most pessimistic predictions of sea level rise—may not obtain economic, political or community approvals. Coastal engineering and design require agreement on criteria, including assessment of risks and extent of future flooding.

The Thames Estuary 2100 Plan [2] was developed in response to the historic flooding of the Thames River, especially the high off-season floods in 2000 and 2008. The plan, abbreviated "TE2100," was the first major flood-risk project in the United Kingdom to adopt climate change adaptation in its planning perspective. The Report takes a long view, "…to address the issues in the context of a changing climate and varying socio-economic conditions that may develop over the next 100 years" [p. 29].

The TE2100 plan defines a "managed adaptive approach", with checkpoints to implement corrective measures appropriate to new conditions as they may develop (Fig. 4.2).

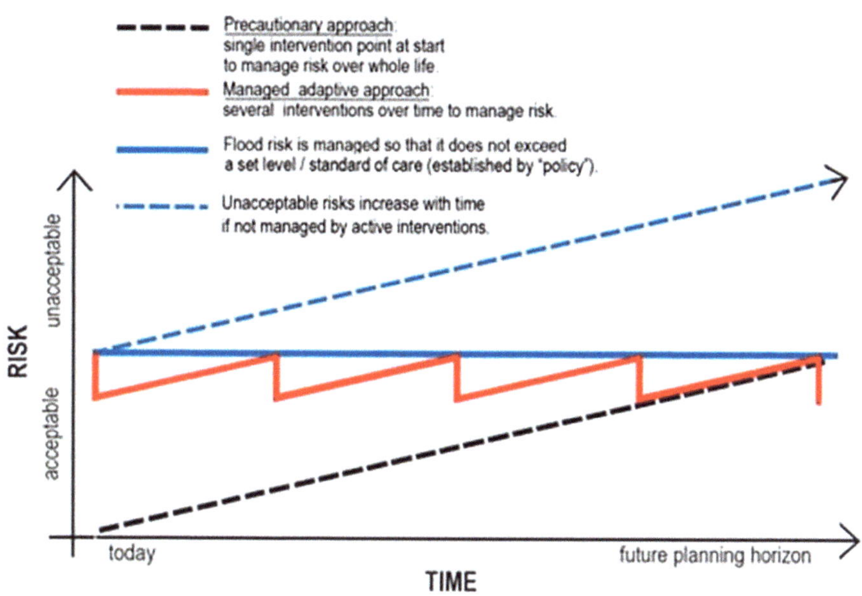

Fig. 4.2 Managing flood risk through the century using the TE2100 managed adaptive approach [2]

A long-term planning perspective is required of any project that is of a scale that responds to climate change projection. The footprint of coastal developments lasts well over a century. The TE2100 plan establishes a process of continued assessment of variables, including Mean Sea Level, peak surge and river flows, condition, and reliability of flood defenses, changes in land development and values, extent of erosion/deposition, and public and institutional attitudes to risk. Three time periods over 100 years are projected: the first 25 years, the mid-century (2030–2049), and end of century (2050–2100), recognizing, "by 2070, flood risk management arrangements must be in place to take us to the end of the century—and beyond" [p. 40].

(2) *Plan with a range of options*
Coastal adaptation planning should evaluate the benefit and costs of a range of options to provide flexibility in initial and longer term adaptive management. Many options are mutually supportive, provide redundancy, and increase the safety factors of an entire set of adaptation measures.

Adapting to Climate Change: A Planning Guide for State Coastal Managers [3] summarizes climate change phenomena that define practices for coastal adaptation. The guide reviews coastal adaptation measures in terms appropriate to the uncertainties, scale and financial commitments required of coastal adaption, including [p. 53]:

- **Win–win options**—Cost-effective adaptation measures that minimize climate risks or exploit potential opportunities but also have other social, environmental, or economic benefits. In this context, win–win options are often associated with those measures or activities that address climate impacts but which also contribute to climate change mitigation or meet other social and environmental objectives.
- **No-regrets options**—Cost-effective adaptation measures that are worthwhile (i.e., they yield net socioeconomic benefits) whatever the extent of future climate change. These types of measures include those that are justified (cost-effective) under current climate conditions (including those addressing its variability and extremes) and are also consistent with addressing risks associated with projected climate changes.
- **Low-regrets (or limited-regrets) options**—Adaptation measures where the associated costs are relatively low and where the benefits, although mainly met under projected future climate change, may be relatively large.
- **Flexible adaptation options**—Measures designed with the capacity to be modified at a future date as climate changes.

(3) *Address social, economic, and environmental benefits*
A broadly inclusive set of economic, social, and environmental goals enlarges community support, supporting adaptation measures that achieve multiple benefits.

Adapting to Coastal Climate Change: A Guidebook for Development Planners [4] lists a diverse set of goals (Table 4.1) in terms of climate adaptation and societal benefits including:

(1) Maintain functioning and healthy coastal ecosystems
(2) Reduce exposure and vulnerability of the built environment
(3) Maintain livelihood opportunities and diversify options
(4) Reduce risks to human health and safety
(5) Strengthen governance frameworks for coastal adaptation.

(4) *Integrate multidisciplinary knowledge and outcomes*
Coastal adaptation requires knowledge and analytic approaches of many disciplines, including climate science, hydrology, ecological, and conservation biology, combined with the engineering, planning and design and construction disciplines.

NOAA Coastal Planning Advisor [5] describes how coastal adaptation incorporates multidisciplinary integration of planning issues. The NOAA Coastal Planning website provides updates on the development of guidelines on each of the following topics:

- Aquaculture
- Climate Change
- Energy and Government Facility
- Land Use Planning
- Marine Debris
- Nonpoint Source Pollution
- Oceans and Great Lakes Planning
- Public Access
- Erosion and Shoreline Protection
- Special Area Management Plans
- Wetland Conservation and Restoration.

As one example, the *Coastal Planning Advisor* offers guidance for marine debris:

To PLAN FOR MARINE DEBRIS Describe the Coastal Issue and Desired Outcome(s)

1. *What are the sources of land-based marine debris in the area?*
2. *What amount of land-based marine debris has accumulated or currently accumulates on shorelines, in benthic habitats, or in pelagic waters that are neither close to the bottom nor near the shore?*
3. *What are the impacts of water-based marine debris on shorelines, in benthic habitats, and in pelagic waters in the area (e.g., entanglement, ingestion, ghost fishing, habitat damage, navigation hazard, human health and safety risk, aesthetics, and economic)?*
4. *What are the sources of water-based marine debris in the area (e.g., solid waste; lost cargo; abandoned vessels; or abandoned, lost, or otherwise discarded fishing gear)?*

Table 4.1 Adaptation goals and climate change impacts [4]

	Goal	climate change impact
1	Healthy coastal ecosystems	
1.1	Coastal wetland protection restoration	Buffers against extreme events. storm surge, erosion and floods: limits salt water intrusion
1.2	Marine conservation agreements	Improves ecosystem resilience improves economic and social conditions
1.3	Marine protected areas	Maintains healthy habitats and fisheries productivity. acts a marine "refitgia"
1.4	Payment for environmental services	Provides incentives to protect critical habitats, against damages from flooding, surge, erosion
2	Built environment is less exposed	
2.1	Beach and dune nourishment	Protects shores, beaches: serves as "soft" buffer against flooding, erosion, scour, water damage
2.2	Building standards	Reduces damages and human safety risks from extreme events, sea level rise, and flooding
2.3	Coastal development setbacks	Reduces infrastructure losses and human safety risks of sea level rise, storm surge, and erosion
2.4	Living shorelines	Mitigates erosion, protects people and ecosystems and natural shoreline variability
2.5	Structural shoreline stabilization	Temporary buffer against erosion and flooding from sea level rise, strom surge, wave action
3	Diversified livelihoods	
3.1	Fisheries sector good practices	Helps sustain livelihoods, food security and marine biodiversity against climate impacts
3.2	Mariculture Best management practices	Helps to mitigate extreme climate, acidification, sea level rise and sea surface warming
3.3	Tourism best management practices	Helps to promote sector sustainability and safeguard against inappropriate uses
4	Enhanced human health/safety	
4.1	Community-based disaster risk reduction	Increases local community resilience and ability to respond to the effects of extreme climate events
4.2	Hood hazard mapping	Informs coastal planning and policy, to mitigate and adapt to the impact of floating
5	Overarching planning governance	
5.1	Coastal watershed management	Preserves estuaries as storm buffers and protect against coastal groundwater salinization
5.2	Integrated coastal management	Provides process that defines goals, actions to address costal issues, including climate change
5.3	Special area management planning	Improves management of discreet geographic areas with complex coastal management issues

5. *Are there certain areas where the impacts of marine debris are greatest (land and ocean)?*
6. *What are the challenges in addressing marine debris?*

(5) *Combine green and blue agendas*
Construction of buildings and infrastructure that is resistant to climate change includes fortifying structures for extreme weather. At the same time, coastal adaptation projects should capture the opportunity to increase the energy efficiency, durability, health, and productivity benefits of "green" design.

Achieving Hazard-Resilient Coastal and Waterfront Smart Growth [6] lists widely promoted "Smart Growth" strategies alongside hazard mitigation strategies, indicating benefits as well as challenges of relocation and flood protection of traditional coastal communities (Table 4.2).

Table 4.2 Interaction between smart growth principles and natural hazard mitigation

NATURAL HAZARD MITIGATION STRATEGIES	COMMON BENEFITS + & CHALLENGES −	SMART GROWTH STRATEGIES
Use green infrastructure to manage stormwater	*+ mitigates flooding, climate* *+ reduces stormwater flow* *+ provides open space*	Encourage green infrastructure at all scales
Protect hazard-prone areas along the shore	*+ protects lives & property* *+ provides public shoreline & recreational areas*	Create walkable /bikable access along waterways for public uses
Protect environmentally sensitive areas	*+ enhances ecosystem values & services* *+ provides open spaces*	Protect critical natural areas, wetland, woodlands
Plan emergency access & evacuation routes	*+ reduces traffic congestion* *+ increases public safety*	Provide transportation options including public transit
Elevate / floodproof buildings & infrastructure	*+ mitigates flooding, climate* *− requires attention to community image, access*	Preserve / enhance sense of place, cultural, heritage, equity
Relocate development out of hazard-prone areas	*− existing transit RR and coastal hwys in floodplain* *− relocation may discourage coastal revitalization*	Direct development to compact, transit nodes & infrastructure

The *top* four strategies complement and support one another. The *bottom* two could conflict subject to planning and design [6]

(6) *Plan for impacts from coastal water to upland communities*
Any coastal location has impact on its coastal neighbors, e.g., one project site can help or hinder water and wave action along the shore. Analysis of local hydrology is a key to present and future flooding and debris flow impacts. Additionally, vital connections of natural and built systems need to be maintained from inland to shore, to enhance continuity of natural systems of streams, vegetation, and hydrology of roads, infrastructure, and adjacent buildings.

Adapting to Climate Change: A Planning Guide for State Coastal Managers [3] details climate change impacts and risk to coastal regions including: rising sea levels, storm intensity/frequency, changing precipitation patterns, increasing air temperature, increasing water temperature, and ocean acidification. The guide cites the NOAA's Coastal Planning website as an updated source of data—elevation, land cover, shoreline, benthic and orthographic imagery, socioeconomics—and related resources, including the *Coastal Inundation Toolkit* with custom visualization tools, county-level data, and guides for creating local inundation maps.

Coastal Climate Resilience Urban Waterfront Adaptive Strategies [7] describes issues and opportunities addressed by coastal adaptation including [p. 117]:

- Vulnerable populations
- Types of buildings, both in terms of their use and structural characteristics
- Critical facilities and infrastructure
- Parks and open spaces
- Ecological systems
- Potentially hazardous materials and uses.

The NYC Report then defines as "reach strategies," a set of planning projects at large scale required to have major impacts beyond the individual project:

Upland Strategies: Strategies that do not involve direct impact on the water or the shoreline, but involve changes to areas inland of the shoreline.
Shoreline Strategies: Strategies to armor or reinforce the shoreline to protect from erosion, block storm surge, or attenuate waves.
In-Water Strategies: Primarily deployed seaward of the shoreline to protect upland areas from erosion and wave forces by attenuating waves, or to reduce the height of storm surge [p. 64].

The report offers a typology of coastal areas to represent a range of geomorphology and land use conditions (Fig. 4.3).

(7) *Plan for living shoreline ecological zones*
The term "living shorelines" refers to coastal adaptation measures that mitigate wave and impacts, erosion, and debris flow, while also providing ecological functions and natural habitat. Measures may include offshore breakwaters, marsh enhancement, dune vegetation, as well as fragmented barriers that provide for vegetation and marine life. Coastal estuaries and river environments are dependent upon the flux of salt water and freshwater to support their

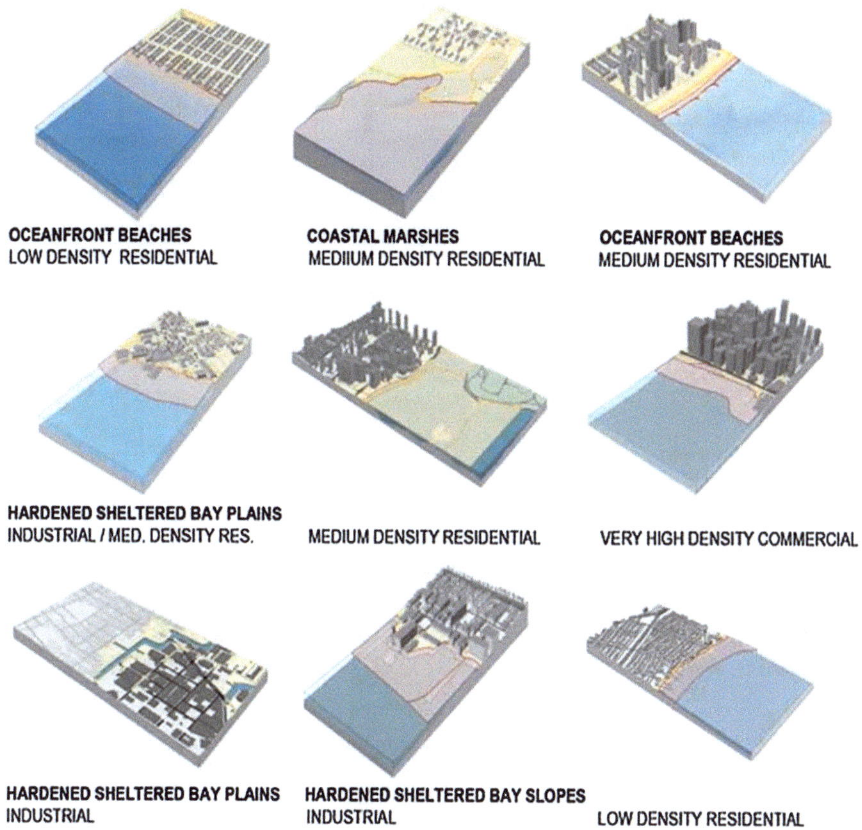

Fig. 4.3 Coastal area typologies. Adapted from NYC department of city planning's coastal climate resilience urban waterfront adaptive strategies (2013)

rich diversity of habitat that sustains fish and bird populations, so that water flow has to be maintained between land, rivers, and oceans. Coastal adaptation has a unique opportunity to provide space for future advancement of these unique ecological zones, as existing coastal areas are covered by rising sea levels.

Engineered Approaches for Limiting Erosion along Sheltered Shorelines [8] provides a review of measures for coastal mitigation and adaptation, based on data developed for the New York State Department of Environmental Conservation's Hudson River Sustainable Shorelines Project. The report gives a detailed assessment of living shoreline measures, compared in terms of construction and maintenance costs, including "adaptability" to advancing sea levels (Table 4.3).

Marshes on the Move [9] demonstrates analysis methods and models to evaluate a coastal area to allow for marsh "advancement" landward. A defining principle of

Table 4.3 Representative costs and benefits living shoreline practices [8]

	APPROACH (Soft–Hard)	CONSTRUCTION COST (Low–High)	MAINTENANCE COST (Low–High)	ADAPTABILITY (Low–High)
Bulkhead	■■■■■	■■■□□	■□□□□	■□□□□
Gabions	■■■■□	■■■□□	■■■□□	■■■■□
Revetments	■■■■□	■■■■■	■■□□□	■■■■□
Levees	■■■□□	■■■■■	■■□□□	■□□□□
Groins	■■■□□	■■■■□	■■□□□	■■■□□
Rip-rap	■■■□□	■■□□□	■□□□□	■■■■■
Green walls	■■■■□	■■■■□	■■□□□	■□□□□
Timber Cribbing	■■■■□	■■■□□	■■■□□	■■■□□
Geotextile Roll	■■■□□	■■■□□	■■□□□	■□□□□
Vegetated Geogrid	■■■□□	■■■■□	■■■□□	■■■□□
Coconut Fiber Rolls	■□□□□	■□□□□	■■□□□	■□□□□
Reed Clumps	■□□□□	■□□□□	■■■□□	■■■■□
Breakwater	■■■□□	■■■■■	■□□□□	■■■■□
Floating Breakwater	■■□□□	■■■■■	■■■□□	■■■■□
Living Reef Breakwater	■□□□□	■■■□□	■■□□□	■■□□□

resilient coastal adaptation is to the extent possible to reserve near-coastal areas to preserve and extend the functions of natural ecosystems, such as marshes, wetlands, and estuaries, as sea levels rise (Fig. 4.4).

- Rate of Sea Level Rise: Not only the amount but the rate of sea level rise is a key factor.
- Tides: Tide levels are important in determining wetland extent and persistence.
- Salinity: If the salinity regime changes, vegetation, and wetland functioning may change.
- Elevation: Elevation is one of the most important data components for modeling sea level rise.
- Sediment Dynamics: If sediment accumulates, or accretes, on the wetland surface as fast as the sea level rises, then the wetland may avoid being submerged under the rising sea.

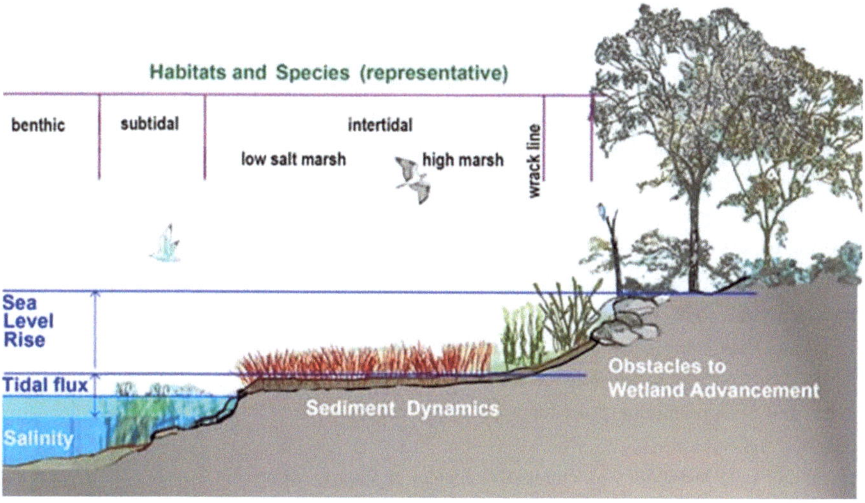

Fig. 4.4 Key parameters for wetland migration analysis in response to sea level rise [9]

- Habitats and Species: Land cover data and habitat change rules can be used to project the effects of sea level rise on the locations of habitats and species. Obstacles to marsh migration are an important consideration when modeling these changes [p. 9].

(8) *Evaluate site infrastructure and building suitability*

A thorough approach to site analysis is required in coastal site selection and design. Maps of flooding that may determine regulatory requirements represent minimum requirements only, insufficient to indicate risks. Mapped data may not represent recent development, erosion, or sea level rise vulnerability. The site analysis process includes documentation of historical and future development alterations, regional hydrology and other climate trends, geology and soils, including the local bathymetry (underwater coastal and ocean floor).

FEMA Coastal Construction Manual [10] is the authoritative reference for coastal construction in the U.S. Its findings are based on building damage assessment and expert recommendations following major catastrophic storms that have impacted the United States and territories over the past 50 years. It represents the guidance for the U.S. National Flood Insurance Program (NFIP). It is regularly updated by technical bulletins and advisories.

The Manual provides detailed guidance on design for coastal construction, including siting, erosion control, building design, and construction. Table 4.4 is representative of items that the *Coastal Construction Manual* lists for property and site selection.

Table 4.4 Checklist to evaluate coastal property [10] [p. 4–7]

Planning and regulatory information
Local and State Natural Hazard Mitigation Plans (NHMPs)
Land use designation at property and adjacent properties
Zoning classification and resulting restrictions on use
Building codes and local amendments
Flood hazard are: elevation and construction requirements
Erosion hazard area: construction setbacks and regulations
Natural resource protection area: siting, construction, or use restrictions
Easements and rights-of-way on property (including beach access locations)
Local and state siting, erosion control and construction regulations
Regulatory front, back, and sideline setbacks
Local and State permitting procedures and requirements
Riparian rights
Local and state restrictions on cumulative repairs or improvements
Subdivision plat and homeowners association requirements and restrictions
Hazard disclosure requirements for property transfer
Physical and natural characteristics
Soils, geology, and vegetation, site, local area, regional
Shoreline land and nearshore (benthic) topography/morphology
Site drainage (surface and groundwater)
Littoral sediment supply and sediment budget
Erodability of nearshore bottom
Erosion contract structures—typo, age, condition, history
Proximity to inlets and navigation structures
Previous or planned community/beach/dune restoration projects
Relative sea level/water change levels—land subsidence or uplift
Sea Level Rise projected by future probable conditions
Protection and extension (advancement) of natural coastal buffers
Infrastructure and supporting development
Emergency access, fire and police life safety and property protection
Evacuation routes and/or local community safe locations
Electric, gas, water, communication/utility linos, shut-offs and hook ups
Sewer, septic, sanitary requirements and health provisions
Debris and debris refugia ***i*** impoundments in storm and surge conditions
Financial considerations
Intended uses—owner occupied and/or rental, private or public
Real estate taxes and property market values
Development, permit and management fees
Multihazard risk insurance: availability, premiums, deductibles, exclusions
Community association assessments (present and future liabilities)

(continued)

Table 4.4 (continued)

Maintenance and repair (present and future)
Building damage (insured and uninsured) (prior, present, and future)
Owner/insurance requirements for safety and property risk protection
Adaptability to future probable conditions (climate, economic)

Design for Flooding [11] defines design and construction practices that build upon FEMA and other coastal adaptation guidance, including changing site conditions. In Fig. 4.5, an assumed 3 ft. Sea Level Rise (SLR) is indicated to illustrate changes in designated flood zones and resulting risks including the scenario that:

(1) Flooding is extended landward;
(2) Wave heights increase due to increased depth of water and wider fetch of open water exposed to wind and wave build up;
(3) Saltwater intrusion and threat is extended landward to freshwater aquifers; and
(4) Surge height and breath is increased due to greater ocean water volume and fetch.

Coastal Flood Hazard Zones (U.S. FEMA definitions)

V	*Velocity zone, subject to wave height greater than 3 ft*
Coastal A	*Subject to flood and water height between 3 ft and 1.5 ft*
A Zone	*Subject to inundation, wave height less than 1.5 ft*
X Zone	*Outside Special Flood Hazard Area, possible flooding based on local conditions.*

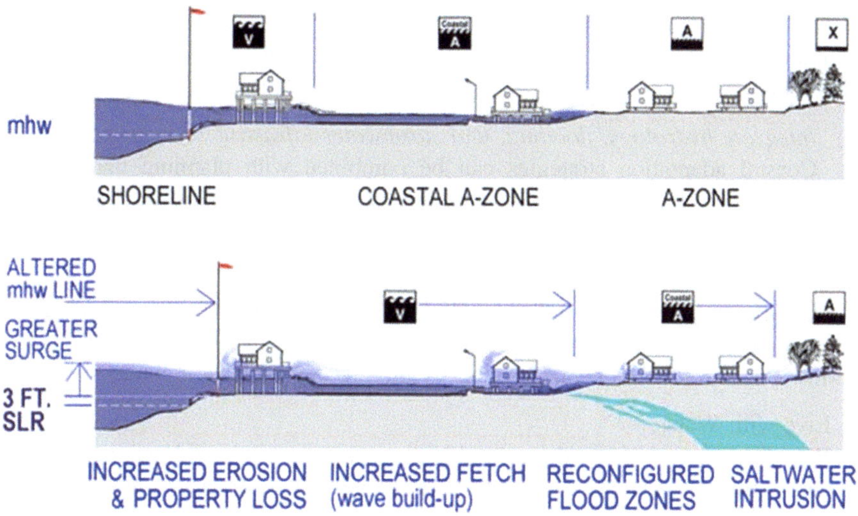

Fig. 4.5 Shoreline transect: flood zones and risks are extended landward with sea level rise [11]

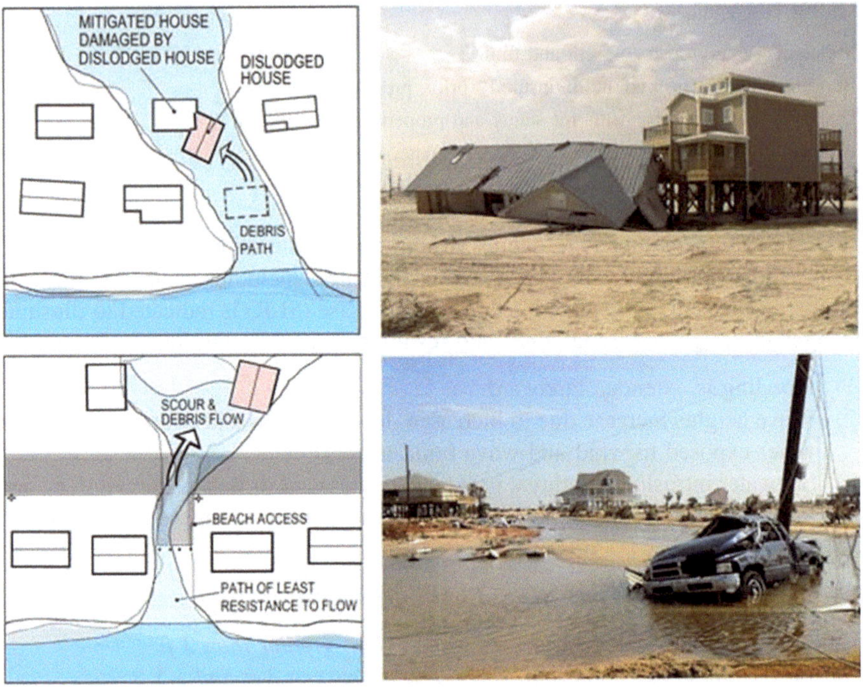

Fig. 4.6 Debris flow damage typical of coastal storms. The diagram on the *top* shows a debris path dislodging a house while the diagram on the *bottom* shows scour and debris blocking flow and disrupting beach access [11]

Coastal adaptation requires consideration of measures at a scale well beyond any one project, property, or parcel. This point is illustrated by examples of community-scale impacts, where the weakest link determines community-scale vulnerabilities (Fig. 4.6).

(9) *Integrate hydrology, flooding, and stormwater infrastructure*
Coastal adaptation strategies can be combined with planning that protects water management systems, including precipitation, stormwater flow, and aquifer replenishment and cleansing, essentially combining "green, blue and grey" approaches.

The Greater New Orleans Urban Water Plan [11] presents planning principles that embrace a regional scale design for water management, including flood water, stormwater, surface water, and ground water. These principles include:

- Live with Water:
  Water is a fact of life on the delta. Making space for water and making it visible across the urban landscape allows it once again to be an asset to the region.

- Slow and Store:
  Stormwater moving fast is hard to manage. Holding it where it falls, slowing the flow of water across the landscape, and storing large volumes of rainfall for infiltration and other uses are fundamental strategies for managing stormwater. Pump stations are activated when necessary, rather than as a default every time it rains.
- Circulate and Recharge:
  Surface waters and groundwater move naturally across and within every delta. Incorporating surface water flows and higher water levels into everyday water management improves groundwater balance, water quality, and the region's ecological health.
- Work with Nature:
  The region's diverse flora and fauna already store, filter, and grow with water. Integrating these natural processes with mechanical systems enhances the function, beauty, and resilience of the region's water infrastructure and landscape.
- Design for Adaptation:
  Change is constant on the delta. Designing systems for dynamic conditions, and to support diverse uses, economic development, and environmental restoration maximizes the value of necessary water infrastructure investments.

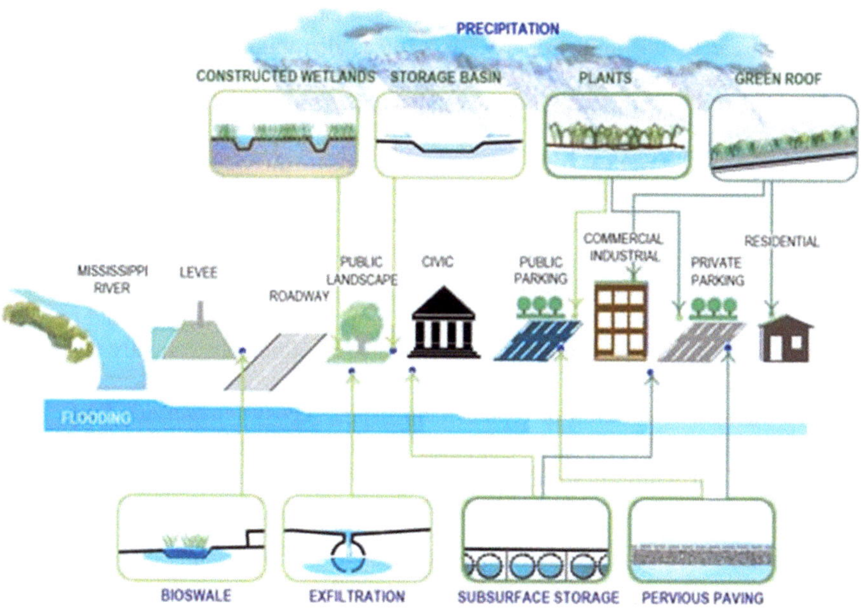

Fig. 4.7 Integrated natural and engineered system practices represented in the New Orleans urban water plan [12]

- Work Together:
  Water knows no boundaries. Collaborations across neighborhood, cultural, and political boundaries and developing solutions at all scales—from individual properties to regional networks—are prerequisites for building a stronger future [p. 28].

The New Orleans Urban Water Plan is conceived at metropolitan region scale. It includes flood control, soil conservation, natural, and constructed infrastructure to accommodate the region's hydrological assets, defined as "the water assignment": (Fig. 4.7)

> …the term water assignment denotes the volume of stormwater for a given rain event (typically 10-year storm) that exceeds the total storage and pumping capacity of a catchment area. In order to reduce flooding, one of the primary objectives of the Urban Water Plan is to find efficient means for meeting the water assignment for each catchment area. While increasing pumping capacity is appropriate in some instances, reducing runoff and integrating storage features into existing drainage systems provide more cost-effective and sustainable alternatives for achieving the same objective [p. 64].

Summary

Climate change presents risks to coastal zones vulnerable to severe storms, warming waters, pollution, sea level rise, and acidification. Storm surge has the greatest threat to life and property, as well as disruption of established natural ecosystems. Sea level rise adds uncertainty and complexity, entirely dependent upon the rate of global climate change and on-going coastal development. Given a constantly shifting landscape, coastal adaptation has to respond to three unique challenges:

(1) To plan for uncertain future conditions, accounting for risks that may be predicted as probable or possible, but without full reliability, validation or definition in regulatory practices.
(2) To plan for extreme storm events, surge, wind, and wave heights, and resulting debris flow, pollution, and infrastructure losses, which pose equally destructive risks.
(3) To plan for restoration or replacement of the "ecosystem services" of coastal environments that are at risk in order to sustain unique habitats for fish, birds, and animal wildlife, along with the recreational and commercial enterprises that they support.

There are reasoned responses to each challenge and uncertainty can be built into a planning process with corrective measures ready to be phased in over the long term. Extreme impacts of storms, wind and waves can be addressed by more stringent standards in engineering and construction, along with vigilant maintenance of coastal facilities and infrastructure. Preservation and replacement of coastal ecosystems can be made part of long-range planning regulations and best practices.

To summarize, the nine working principles, illustrated by the previous references cited in this chapter, include:

(1) Plan for uncertainty
(2) Plan with a range of options
(3) Address social, economic, and environmental benefits
(4) Integrate multidisciplinary knowledge and outcomes
(5) Combine green and blue agendas
(6) Plan for impacts from coastal water to upland communities
(7) Plan for living shoreline ecological zones
(8) Evaluate site infrastructure and building suitability
(9) Integrate hydrology, flooding, and stormwater infrastructure.

An overarching guideline for coastal adaptation is that at each and every opportunity, ***coastal development should seek to protect, replicate, and improve the ecosystems necessary for the lasting protection of natural and community assets, for present and probable future conditions including climate change***.

Viable solutions are possible when undertaken with long-term commitment to the highest standards of health, safety, and community well-being, while protecting and enhancing natural systems, goals. This guiding vision for a truly sustaining future, demonstrated by examples in the references, is the true motivation and reward of planning for coastal adaptation.

References

1. IPCC (1995) Intergovernmental panel on climate change second assessment. World Meteorological Organization, Geneva. URL: http://www.ipcc.ch/pdf/climate-changes-1995
2. Environment Agency (2009) Thames estuary TE100 Plan: managing flood risk through London and the Thames estuary. United Kingdom Environmental Agency, London, 230 p. URL: www.environment.agency.gov.uk
3. NOAA (2010) Adapting to climate change: a planning guide for state coastal managers. NOAA Office of Ocean and Coastal Resource Management. URL: http://coastalmanagement.noaa.gov/climate/adaptation.html
4. US AID (2009) Adapting to coastal climate change: a guidebook for development planners. U.S. Agency for International Development, Washington, DC, 164 p. URL: http://www.crc.uri.edu/download/CoastalAdaptationGuide.pdf
5. NOAA (2007) NOAA coastal planning advisor. Digital coast, National Oceanic and Atmospheric Administration Office for Coastal Management, Charleston, SC. URL: http://www.csc.noaa.gov/coastalplanningadvisor/#/
6. NOAA (2011) Achieving hazard-resilient coastal and waterfront smart growth. Roundtable report: coastal and waterfront smart growth and hazard mitigation, National Oceanic and Atmospheric Administration, Washington, DC, 26 p. URL: www.coastalsmartgrowth.noaa.gov
7. NYC (2013) Coastal climate resilience: urban waterfront adaptive strategies. Department of City Planning, New York. 134 p. URL: www.nyc.gov/uwas

8. Rella AJ, Miller JK (2012) Engineered approaches for limiting erosion along sheltered shorelines: a review of existing methods. Prepared for the Hudson River Valley Greenway and Hudson River National Estuarine Research Reserve, 2012. URL: http://www.dec.ny.gov/docs/remediation_hudson_pdf/limiteros.pdf
9. TNC/NOAA (2011) Marshes on the move: a manager's guide to understanding and using model results depicting potential impacts of sea level rise on coastal wetlands. The Nature Conservancy, Narragansett, RI, and National Oceanic and Atmospheric Administration, Washington, DC, 24 p. URL: http://northeastoceancouncil.org/marshmigration/
10. FEMA (2011) Coastal construction manual: principles and practices of planning, siting, designing, construction, and maintaining residential buildings in coastal areas. In: FEMA-55 August 2011, 4th edn. Federal Emergency Management Administration, Washington, DC, 253 p. URL: www.fema.gov/media-library/assets/documents/3293
11. Watson D, Adams M (2011) Design for flooding: architecture, landscape and urban design for resilience to climate change. Wiley, New York, 298 p. URL: http://www.wiley.com/WileyCDA/WileyTitle/productCd-0470475641
12. Waggonner D, Ball M (2013) Greater new orleans urban water plan: vol. 1. vision. Greater New Orleans Inc. Regional Economic Development, Waggoner & Ball Architects, New Orleans, LA, 203 p. URL: http://livingwithwater.com

Author Biography

Donald Watson is an architect and author, and former professor and dean of the School of Architecture, Rensselaer Polytechnic Institute. He received the 2002 ACSA Distinguished Professor Award and the 2005 ARCC Haecker Leadership Award for Architectural Research. His publications include *Climatic Building Design* (McGraw-Hill 1983), recipient of the American Publishers Award for Professional and Scholarly Excellence, and *Design for Flooding* (Wiley 2011), coauthored with Michele Adams, P.E. He is editor-in-chief of the McGraw-Hill TimeSaver Standards Series for *Urban Design* (2001) and FEMA Subject Matter Expert on building technology and community design. He is principal of EarthRise design in Trumbull, CT, providing planning and design services for civic, institutional and urban projects, serving governmental and corporate clients in the United States and abroad, with focus upon environmental design, sustainability, and resiliency.

http://nhma.info/uploads/board/Don_Watson_Bio.pdf.

http://execed.gsd.harvard.edu/people/donald-watson.

Chapter 5
Case Study: Hunts Point, the Brooklyn Greenway, and the East Shore of Staten Island

Tricia Martin, Andrew Moore, Richard Roark and Eric Rothstein

T. Martin (✉)
WE Design Landscape Architecture, 33 Flatbush Avenue, 6th Floor, Brooklyn, NY 11217, USA
e-mail: tmartin@wedesign-nyc.com

A. Moore
Quennell Rothschild & Partners, 132 Nassau Street, Suite 1300, New York, NY 10038, USA
e-mail: moore@qrpartners.com

R. Roark
OLIN, 150 S Independence Mall W, Suite 1123, Philadelphia, PA 19106, USA
e-mail: rroark@theolinstudio.com

E. Rothstein
eDesign Dynamics, 402 W. 40th Street, New York, NY 10018, USA
e-mail: ERothstein@edesigndynamics.com

M. Johnson and A. Bayley (eds.), *Coastal Change, Ocean Conservation and Resilient Communities*, DOI 10.1007/978-3-319-41914-5_5

Introduction

Many cities around the world are currently facing the environmental, economic, and social effects of climate change, including an increase in energy costs and demands, internal flooding and storm and tidal flooding. Storms like Katrina, Sandy, and Typhoon Haiyan have resulted in redirected priorities of communities across the world towards integrating storm protection and other climate change adaptation strategies into ongoing planning and design efforts. While there is not a silver bullet solution for mitigating these climate change effects, many cities are adopting plans that are multi-objective, meeting many of their environmental, economic, social, and infrastructural goals.

The densely packed cities of the east coast are caught between two intersecting dilemmas: aging infrastructure and climate change. In the coming decades, cities will be scrambling for funds to invest in new solutions to these chronic problems and the results will have a lasting impact on economic productivity, health, and environmental quality of the region. Following the devastating damage from Hurricane Sandy in 2012, New York City intends to invest billions of dollars on shoreline protections.

The protection measures offer both opportunities and challenges: how does a massive investment with the primary goal of preventing devastation from storm surge impact shoreline programming, habitat, and community's relationship with the water's edge? In this chapter, we discuss three shoreline protection projects in various stages of planning and design. Hunts Point is in the South Bronx, the Brooklyn Greenway extends through multiple shoreline communities in Brooklyn, and the East Shore of Staten Island—one of the hardest hit areas within the five boroughs of New York during Hurricane Sandy.

Hunts Point Case Study

The Food Distribution Center of New York in the South Bronx neighborhood of Hunts Point is confronting the challenges of an aging complex of buildings located in a low-lying peninsula of the East River Estuary.

In 2013, Hunts Point became a focus of study for the PennDesign/OLIN Rebuild by Design team, as part of a presidential initiative to develop innovative resilient projects to protect coastal cities in the wake of 2012s Hurricane Sandy. The team identified the neighborhood of Hunts Point as a focus for resilient design because of the neighborhood's environmental and economic burdens, its position as the hub of New York's food supply and its vulnerability to climate change.

The Hunts Point Food Distribution Center (FDC) is an industry with annual revenues of $5 billion dollars and provides annual wages of $285 million dollars [1] Second, the distribution center is a critical facility for the safe supply of food for New York providing food throughout the region to an estimated 22 million people [2].

But today, nearly a third of this vital complex is in the 100 year floodplain, by 2050 according to New York City sea level rise projections, the entire complex will have a 1 % annual chance of flooding. The neighborhood community is equally under threat suffering from numerous environmental burdens and severe under employment. New York's community health report identified Hunts Point and the adjoining neighborhood of Mott Haven as having the highest rates of asthma rates, cancer, diabetes and premature death in the city of New York [3].

While the flooding threat from the East River was well-understood, the infrastructure commitment to protecting miles of urban coastline required a careful weighing of both costs and benefits. An assessment of both building-based and site-based flood protection options were considered. A site wide strategy was determined to be most advantageous given the age of the facilities and complications with integrating flood barriers adjacent to loading docks and truck yard logistics for food shipments. A perimeter site-based approach was determined to be less disruptive to existing operations at the Food Distribution Center. Significant community benefit would also be realized through the development of a protective greenway connecting the neighborhood to the waterfront, fulfilling the long-standing neighborhood vision of the South Bronx Greenway framework. The design solution envisioned is a combination of levee, seawall and constructed wetland basins.

The flood protection incorporates an applied research model termed "Levee Lab" which proposes to incorporate monitoring and testing of materials that create vegetative substrates to support the ecological function of the estuary and to improve water quality. The conceptual plan includes the creation of salt marsh zones. Salt marshes flood twice daily and are critical ecosystems within the larger estuary. Salt marsh plants, predominantly *Spartina alterniflora* uptake nutrients and metals become sequestered in their root zones. As salt marshes age, their peat layer grows thicker—a result of detritus build-up from annual plant die-back. The peat sequesters contaminants and provides surfaces for bivalves to attach. The bivalves, in turn, help to improve water quality while the peat serves as critical habitat for invertebrates. Fiddler crabs burrow through the peat, creating additional surface area for water treatment. Salt marshes are most commonly noted for their habitat value as fisheries. They serve as foraging and protected ground for small and immature fish. Harbor herons can frequently be seen feeding within salt marshes. Salt marshes are just one example of how Levee Lab seeks to maximize co-benefits into shoreline-based protection.

The concept of Levee Lab was inspired by the varied hydrologic morphology of the East River estuary and Hunts Point's status as a Significant Maritime Industrial Area (SMIA). The peninsula of land that form's Hunts Point is at the confluence of the Bronx River and adjacent to nearby North Brother Island and Riker's Island. Hunts Point's location creates an ideal testing ground for the Levee Lab, given the diverse range of urban shore profiles and hydrologic flows that can be assessed over a relatively short expanse of shoreline.

A preliminary hazard mitigation analysis concluded that the Levee Lab project avoided loss of food inventory, avoided loss of business revenue, avoided downstream impacts of loss of food supply, and would provide over $1.1B in avoided damages. The construction of the flood protection system as envisioned would create an estimated 7000 jobs along the peninsula over an assumed 10 year period of construction. Altogether the project achieved a cost benefit ratio of 2:1 in avoided damages alone. The design and construction of the Levee protection

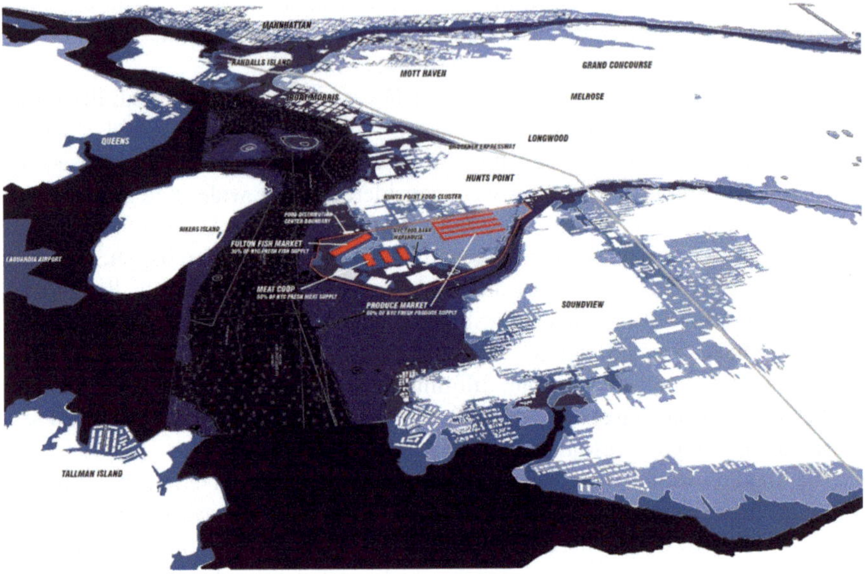

Fig. 5.1 Map of Hunts Point, located in the South Bronx

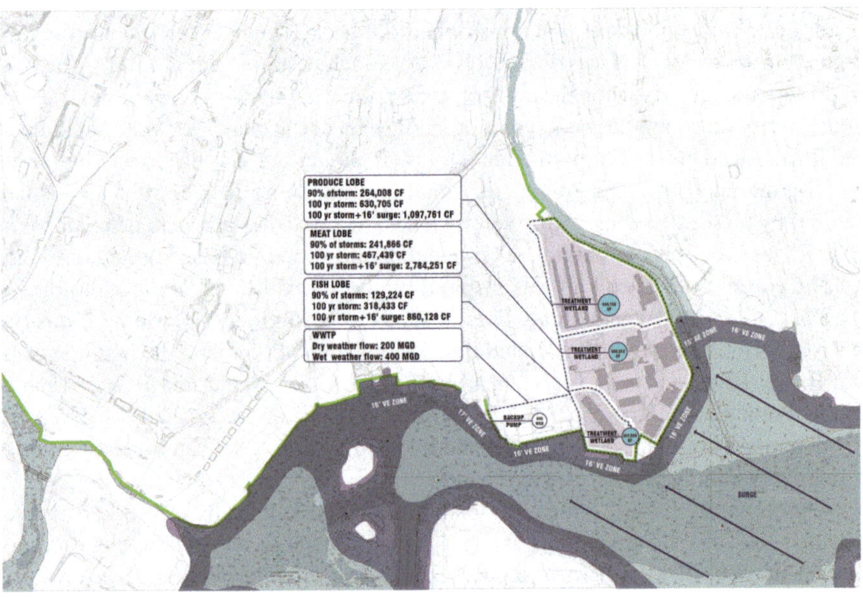

Fig. 5.2 Hunts Point areas or “Lobes” subject to inundation

system will create both temporary and long-term employment for the neighborhood in the form of construction jobs, green jobs in the form of coastal monitoring and upkeep of the greenway, and in the potential for new industry through the manufacture of modular flood protection and locally produced vegetative mats and substrates. As a vision for the future of urban coastal resilience, the project demonstrates planning for a regional infrastructure that is adaptive to change and sensitive to local communities and environments (Figs. 5.1, 5.2, 5.3 and 5.4).

Fig. 5.3 Hunts Point "Levee Lab" infrastructure

Fig. 5.4 Hunts Point reimagined

Brooklyn Greenway Case Study

While the impetus behind the Hunts Point project was shoreline protection, the Brooklyn Greenway initiative was well underway when Sandy hit. Incorporation of shoreline protection was then added to an initiative that had primarily been focused on shoreline access and safe bike and pedestrian transit routes. Greenways enhance, complement and can be central to the planning for resiliency.

Greenways are an effective, multi-objective mechanism for building climate change adaptation strategies into our cities. *The Brooklyn Waterfront Greenway Green Infrastructure Design and Resiliency Guidelines* is a case study demonstrating how greenways contribute to providing open space, recreation, and offer opportunities for alternative modes of transportation, and show how they can be so much more. Forward-looking concerns about changing climate, rising sea level, water quality and coastal habitat enhancements can and should be addressed when designing our greenways. The Brooklyn Waterfront Greenway green infrastructure design guidelines and strategies are intended to provide a coherent expression for Greenway development while addressing a range of issues influencing urban coastlines. Fourteen miles of Brooklyn's waterfront perimeter is to be monumentally transformed by the construction of a continuous corridor for bicycles and pedestrians, triggering the opportunity to integrate forward-looking concerns about changing climate, rising sea levels, water quality, and coastal habitat enhancements. Coastal habitat enhancements may include the restoration of imperiled upland ecosystems such as coastal scrub shrub; construction of fresh water wetlands that manage stormwater while also providing foraging habitat for harbor herons and migratory song birds; salt marshes that filter water and provide critical fishery habitat; and the use of in-water structures such as reef balls that improve fish habitat (Figs. 5.5, 5.6 and 5.7).

Large areas of Brooklyn were flooded by the storm surge generated by Hurricane Sandy in 2012. One year earlier, Hurricane Irene brought heavy precipitation and inland flooding caused by inland runoff. Simultaneously protecting against both types of storm risks is technically challenging—balancing the need to protect properties from storm surge and the need to release stormwater from heavy rains. The Brooklyn Waterfront Greenway will precisely delineate those locations where, one day, Irene will meet Sandy. If these low-lying areas are to survive extreme storm events for the indefinite future, comprehensive interventions are required to mitigate risk to property, and maximize resilience of habitat and infrastructure. For habitat design, this can mean sacrificing some biodiversity by predominantly planting hardy, native species with wide ranges of tolerances for soil and water chemistry. By including design considerations for green stormwater management and storm surge protection, the Brooklyn Greenway presents an opportunity to address these issues in many locations along the route while creating a valuable public amenity.

New York City's Long-Term Control Plan (LTCP), developed to establish compliance with State and Federal environmental regulators enforcing the Clean

Fig. 5.5 Map of the proposed Brooklyn Greenway (*white line*)

Water Act, promises dramatic reductions to Combined Sewer Overflows (CSOs) in the coming decades. The LTCP describes massive infrastructure improvements to the City's sewers, along with Green Infrastructure (GI) practices that manage stormwater through source controls and "green" systems that replicate natural hydrological processes. These practices, however, are largely limited by their disconnection from historic drainage ways and connections for release to open water. Because the Brooklyn Waterfront Greenway will lie over precisely those potential points of discharge, the hope is to demonstrate the benefits of reestablishing the pathways that permit collection, treatment and release of stormwater runoff from

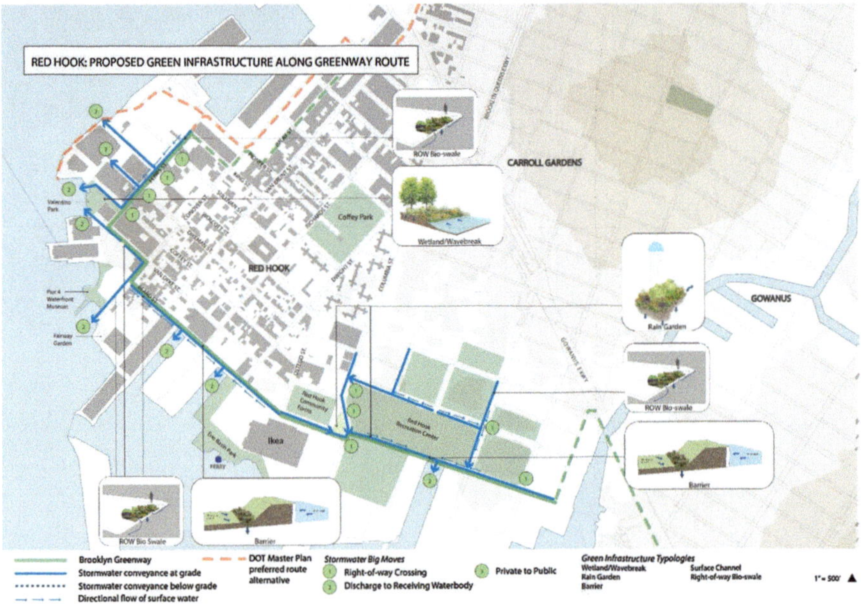

Fig. 5.6 Proposed green infrastructure along the Brooklyn Greenway

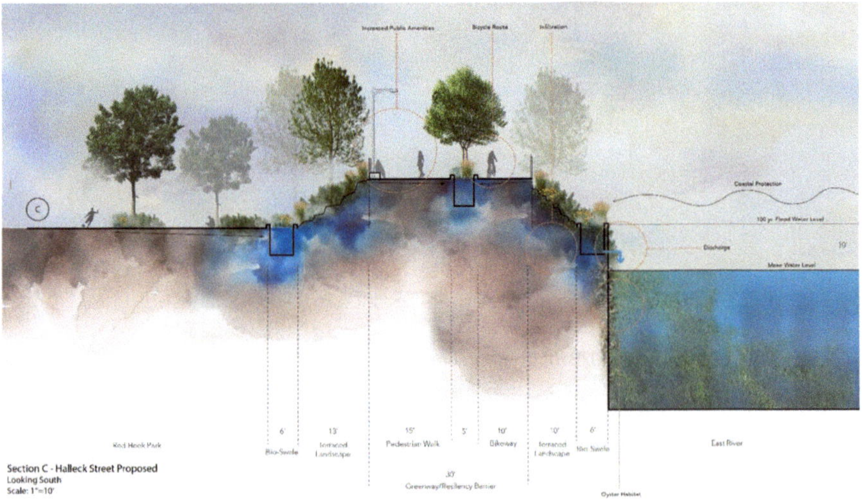

Fig. 5.7 Illustrative section showing street-end interface with levee protection system

developed land to natural waterways. In doing this, the project proposes using many of the tools established for use by New York City's Department of Environmental Protection (NYCDEP) and the Office of Green Infrastructure. However, some new tools are still needed to reach the full potential of GI systems to capture and treat urban stormwater in these coastal applications.

One such tool planned for use by NYCDEP is the high-level storm sewer (HLSS), which redirects runoff collected near shorelines away from the combined sewers toward new points of discharge to the waterways. This form of strategically separating storm and sanitary flows has great potential to reduce CSO, but with minimal use of water quality controls. Once the HLSS becomes common practice in the City, it is expected that State and Federal regulators will impose MS4 obligations to treat stormwater prior to release. The design scenarios in the guidelines document illustrate the use of HLSS' and their potential for integration with living filters and wetland treatment green infrastructure systems. The green infrastructure would focus on improving water quality prior to it reaching the HLSS, while the HLSS will ultimately improve water quality by limiting runoff contributions to the combined sewer system.

East Shore of Staten Island Case Study

Quennell Rothschild & Partners with McLaren Engineers and eDesign Dynamics are currently preparing a master plan for the New York City Department of Parks and Recreation (DPR) for the shoreline parks on Staten Island's East Shore. The master planning follows proposals by the US Army Corp of Engineers (USACE) to construct approximately four miles of armored levee as storm protection for the adjacent low-lying inland areas. NYC DPR is seeking to intelligently alter, augment, or mitigate the impacts of the USACE construction.

The 5-mile waterfront on Staten Island's East Shore has been a destination for beachfront recreation for almost a century, rivaling Coney Island for hotels and beachfront amusements. The gradual disappearance of the resorts and beachfront developments left a conglomeration of City and Federal parks and open spaces. The project site spans approximately from Fort Wadsworth at the foot of the Verrazano Bridge along the neighborhoods of South Beach, Midland Beach, Miller Field, New Dorp Beach, Oakwood Beach, and turning west around the Oakwood Beach Water Pollution Control Plant and terminating inland at Hyland Blvd. Historically susceptible to storm surges, the parkland and adjacent residential neighborhoods along the East Shore saw some of the most extensive damage in the NYC area during Hurricane Sandy in 2012.

In response to the flooding and loss of life during Hurricane Sandy, the USACE is moving forward in planning for the construction of a new Line of Protection (LOP) consisting of armored levees and seawalls built to a level above the elevation that Hurricane Sandy reached. The USACE is obligated to plan for "in kind" replacement of park elements and resources displaced or destroyed by the construction of the LOP. It is anticipated that parts of the LOP will have a promenade to replace the existing boardwalk and walkways along the other portions of the LOP. Access for maintenance and emergency vehicles and pedestrians will be integrated up to and over the LOP.

The USACE will also develop several inland areas on DPR owned lands and on NYC DEP controlled lands, as Interior Drainage Areas that will serve as reservoirs for storm drainage behind the LOP, providing additional protection for the adjacent low-lying residential areas.

Data Gathering for the Shoreline Parks Master Plan

Working for NYC DPR and in conjunction with other city agencies, the QR&P team is performing extensive site analysis, data gathering and designing elements that will integrate the USACE storm barrier into the park land. Studies and data gathering includes inventories of all of the parks resources from ball fields, existing trees, parking lots and structures to the number of benches and light poles along the boardwalk. Interviews are being conducted with the NYC DPR facilities and operations representatives to understand the maintenance and operations of the parks, conditions and demands on the existing infrastructure such as comfort stations, concessions and maintenance facilities, the requirements for lifeguard operations, and standard operating maintenance of the parks and servicing of private concessionaires. Many existing program areas were damaged by Hurricane Sandy and never repaired or rebuilt. The need for repairs and upgrades to the facilities may not be directly impacted by the USACE LOP but the planning of these facilities will be included in the master plans.

The LOP and construction activities will also impact "undeveloped" areas that contain pockets of high value native and naturalized plant species which include coastal shrub communities, and a mature post oak. These areas have been mapped by eDesign Dynamics and investigated with representatives of the Parks Natural Resource Group to identify high value species and the anticipated level of construction impacts.

Impacts of the USACE LOP

The LOP will generally consist of continuous steel sheet piling encased in a levee constructed of geotextile, bedding stone and armor stone ($W50 = 3.0$ ton) and capped with a sandy planting mixture on the ocean side and typically planted with dune grass. Where it occurs, the 38-foot wide elevated promenade walkway will be constructed on the crest of the LOP and on timber piles to the landward side of the levee. The LOP varies in base width from approximately 40–70 feet in plan dimension depending on the elevation of the existing topography and program uses such as access points, or promenade width. The alignment of the LOP encroaches on built structures such as comfort stations and concessions, paths, roadways, sports fields, parking lots, playgrounds, access steps and ramps and locally significant landmarks such as memorials, fountains and plazas. The levee will also eliminate visual access to the ocean from the inland and make the physical access more difficult. In areas where the boardwalk currently exists, one can see the ocean through the boardwalk and pilings and in places ad hoc paths and trails currently lead under the boardwalk directly to the beach. The top of the levee and proposed esplanade will be at an elevation more than 5 feet above the existing boardwalk elevation. Existing facilities including a privately leased restaurant and catering facility, a concession stand, a restored fishing pier and several comfort stations will not physically be disturbed by the construction of the LOP but will be more than five feet below the proposed adjacent promenade making accessibility and sight-lines difficult or impossible to maintain.

The LOP follows along and will impact several natural areas; along with the changes brought by the artificial conversion of low-lying areas into interior drainage storage areas, the LOP will potentially have significant impact on the hydrology of sensitive plant communities even if they are not materially disturbed by the construction. The extent of the USACE alterations to the topographic elevation, impacts to the groundwater and surface water and impacts to the levels of salinity, are unknown at this point.

The Shoreline Parks Master Plan

The Shoreline Parks Master Plan must be completed prior to the USACE developing their final design drawings for the LOP. The Request for Proposals from NYC Parks describes a process for site analysis and data gathering, soliciting public input for the plans, developing an overall conceptual plan for the Shoreline Parks and identifying up to five specific areas for more detailed investigation and design options. The resulting plans are to identify opportunities to influence the USACE plans by proposing modifications to the plans, by identifying required Parks facilities for replacement in kind, and by proposing "betterments" which could be incorporated into the USACE plans at additional cost to the project.

At this writing, the team is developing the initial conceptual plans for presentation to the adjacent communities for their feedback. Some initial proposals that have come out of the work to date are as follows:

- In areas such as Oakwood Beach, where undeveloped areas will be made more easily accessible, natural areas can be planted with native plants that are appropriate to the location and will provide a thriving habitat that can be accessed in a sensitive manner as a park amenity. Habitats include coastal scrub shrub in upland areas. A salt marsh is also being proposed that will be fed with ocean water via a tidal stream.
- Recommendations for the inland drainage areas will be made to construct them in such a way to provide sustainable plant and animal habitat increasing the natural value of the areas. The team is recommending the inclusion of permanent, shallow water in the expansive detention zones to be planted with deep water emergent plants such as pickerelweed, duck potato, bulrush, and arrow head. These plants will provide critical habitat for shore birds, migratory birds, herptiles, and insects. They will also improve water quality. The team is proposing to use hardy, aggressive growing species. By limiting diversity and focusing on hardy native species, the wetlands will be better suited to resist invasion from *Phragmites australis*. The permanent surface water will also help to limit the establishment of *Phragmites* along the wetland edge.
- The shoreline parks were developed at different times in an ad hoc fashion and the DPR park lands span over different districts that complicate the management of the park as a whole. The NYC park land is interrupted by Miller Field, formerly an Army airstrip and facility which is now mostly active recreation park land of 187 acres owned by the National Parks Service. The patchwork character of the park will be addressed by the proposed master plan to reduce inefficiencies or duplication of services and improve linkages. The NYC Department of City Planning is studying the major feeder streets as commercial corridors and the master plan is looking at the possibility of relocating some of the entrances and parking facilities to simplify how the vehicular circulation works and to enhance and clarify the entrances to the parks.
- The built facilities in the park also have been constructed over a period of years and the overall placement and functionality can be improved. There are a few year-around comfort stations in the park despite the presence of year-around visitors. Offices, lifeguard stations, and comfort stations that were damaged by Hurricane Sandy and not repaired or rebuilt will be considered as a longer term issue for the reconstruction of the park.
- Park maintenance and operations, concession service and waste removal, beach grooming and cleaning will be analyzed and organized systemically to try to remove inefficiencies and conflicts with public uses.
- Parking will be studied to see if the amount of paved surface can be reduced by layout efficiencies, reallocation of parking lots and integration of green infrastructure.
- Access points over the LOP will be reconsidered in the context of the park programs. Access points for pedestrians should be logical and effortless in providing access up and over the LOP. Wherever possible, fully accessible paths will be provided at all locations. Vehicular access points for maintenance and

emergency vehicles should be conveniently located and discretely separated from public access.

- The construction of the LOP may provide opportunities to integrate park facilities into the structure of the levee or elevated promenade. There is potential for bleachers overlooking sports fields, gracious steps and seating leading down to the beach, or incorporating comfort station or maintenance structures into the LOP to unite the facilities into the dune landscape.
- In areas such as Oakwood Beach, where undeveloped areas will be made more easily accessible, natural areas can be planted with native plants that are appropriate to the location and will provide a thriving habitat that can be accessed in a sensitive manner as a park amenity.
- Recommendations for the inland drainage areas will be made to construct them in such a way to provide sustainable plant and animal habitat increasing the natural value of the areas.
- The bike path can be formalized throughout the entire park as a transportation resource as well as for recreational opportunities. The park path system can potentially follow the extent of the LOP and connect Fort Wadsworth to the Staten Island Greenbelt White Trail providing an enormous potential for recreational trails extending throughout Staten Island.

References

1. An economic snapshot of The Hunts Point Food Distribution Center Albany, Office of New York State Comptroller (2008). https://www.osc.state.ny.us/reports/economic/huntspoint08.pdf
2. NYC food by the numbers: The Hunts Point Food Distribution Center, New York City Food Policy Center, September 18th (2015). http://nycfoodpolicy.org/nyc-food-numbers-hunts-point-food-distribution-center/
3. Hunts Point and Mott Haven Community Health Profile, New York City Department of Health and Mental Hygiene (2006). http://www.nyc.gov/html/doh/downloads/pdf/data/2006chp-107.pdf

Authors Biography

Tricia Martin, Founding Partner, WE Design Landscape Architecture, a New York City-based design firm specializing in innovative, buildable design solutions for cities, landscapes, and the built environment.

Tricia is a licensed landscape architect and has worked on a variety of projects that include a community action plan for Grand Ravine in Port-Au-Prince, green infrastructure design guidelines for the Brooklyn Greenway, and a sustainable master plan for a private estate in Pedasi, Panama. While working for Mathews Nielsen Landscape Architects she won an Arts Commission Award in 2006 for her work on the American Airlines Flight 587 Memorial in the Rockaways, Queens.

Tricia is an Adjunct Professor at the Graduate School of Architecture, Planning and Preservation at Columbia University where she teaches urban design studios. She is regularly asked to speak about issues related to green infrastructure and landscape architecture. In 2011, Tricia represented the American Society of Landscape Architects during a senate briefing in Washington, DC on the topic of "Using Green Infrastructure Practices to Protect and Restore Clean Water for Communities." Tricia is Past-President of the New York Chapter of the American Society of Landscape Architects, which she served from 2008–2011.

Andrew Moore, Partner, Quennell Rothschild and Partners LLP, is a landscape architect with over 35 years' experience in New York City. He has designed a broad selection of the firm's notable work, including the master plan for Hudson River Park, multiple projects on the campus of Princeton University, the Central Park Children's Zoo, the restoration of historic parks and landscapes in the northeast and residences in New York, Connecticut, Florida, and New Mexico. Projects include collaborations with architects Frank Gehry, Toshiko Mori, Rafael Vinoly, and artists Paula Hayes and Monika Gryzmala.

He has taught at Parsons School of Design, University of Pennsylvania and Washington University. Award winning projects include the ASLA merit award for the Hudson River Park master plan, Historic Preservation Award for Monument Drive, Princeton New Jersey, and the International Masonry Institute award for Scudder Plaza, Princeton University. Mr. Moore holds a BLA from the University of Oregon and is a registered landscape architect in New York and New Jersey.

Richard Roark, Partner, OLIN, RLA, ASLA, LEED AP BD+C work spans a range of environmental scales and typologies, focused on expanding the civic capacity of the landscape. Projects such as the Presidential Sustainability Initiative: Rebuild by Design, the new U.S. Embassy in London; Dilworth Park in Philadelphia, Pennsylvania; exemplify a practice focused on sustainable, economical, and well-crafted public design. His work in sustainability and social resiliency has reimagined investment in neglected neighborhoods and urban landscapes notably through the strategic plan for Detroit's Eastern Market and the award winning Living Cities project Patchwork Philadelphia. He has also recently completed a vision plan for the Philadelphia City Branch abandoned rail line; a plan to reconnect many neighborhoods and cultural institutions to Fairmount Park along the historic elevated City Branch rail line. His most recent published work includes content for the LA River Index, a publicly accessible, online compendium of research and data on the history, current context, and future potential of the Los Angeles River. He has been recognized as an active participant in public design advocacy and as an outstanding volunteer of the Philadelphia Community Design Collaborative (CDC).

Richard began his studies in landscape architecture at Auburn University, where he earned a Bachelor of Environmental Design and then a Master of Landscape Architecture. He first joined OLIN in 2001 upon receiving his Master of Community Planning. He has served as an invited guest juror at Arizona State University, Auburn University, Temple University and University of Pennsylvania, and is a past member of Auburn University's Landscape Advisory Council.

Eric Rothstein, Managing Partner, eDesign Dynamics is a hydrologist with 20 years of experience working on a variety of sustainability projects. Mr. Rothstein's career has focused on ecosystem restoration and water resources planning within urban centers. He currently leads the team investigating the sustainable water resource planning for multiple projects in New York City. His international work includes water resource and ecosystem planning in Nepal, Sao Paulo, Brazil, rural Rwanda, and the Aegean coast of Turkey. Mr. Rothstein previously worked as a project manager for New York City for over 7 years. In that capacity, he managed ecological restorations including salt marshes, fresh water wetlands, grasslands, and forests and developed naturalized stormwater management structures. Examples of Eric Rothstein's built work can be found in all five boroughs of New York City and beyond. He has also worked for a national ecological restoration firm where he worked on master planning of ecological, stormwater management and soil components for large-scale developments. Mr. Rothstein has lectured on wetland restoration and stormwater management issues at the graduate level at Harvard, Columbia, and The University of Pennsylvania. He teaches continuing education courses to professionals through the State University of New York College of Environmental Science and Forestry. Eric Rothstein holds an M.S. in Soil and Water Engineering/Hydrology from Cornell University and a B.S. from the University of Wisconsin, Madison.

Chapter 6
Case Study: Seahurst and Percival

Peter Hummel and Anna Spooner

Seahurst Park and Percival Landing Case Studies

The Seahurst Park Ecosystem Restoration and Percival Landing Major Rehabilitation projects showcase how Anchor QEA tackles complicated shoreline projects to achieve a diverse set of goals. Both projects are located on Washington

P. Hummel (✉) · A. Spooner
Anchor QEA, LLC, 720 Olive Way, Suite 1900, Seattle, WA 98101, USA
e-mail: phummel@anchorqea.com

A. Spooner
e-mail: aspooner@anchorqea.com

M. Johnson and A. Bayley (eds.), *Coastal Change, Ocean Conservation and Resilient Communities*, DOI 10.1007/978-3-319-41914-5_6

State's Puget Sound, the second largest estuary in the United States. They are both heavily used public waterfront parks that balance recreational access with nearshore ecological restoration.

The federal Endangered Species Act (ESA) listing in 1999 of Puget Sound Chinook salmon as "Threatened" and ESA listings of other fish species and marine mammals has provided a funding mechanism for species recovery and ecosystem restoration in Puget Sound. Much emphasis has been placed on shallow water, "nearshore" areas consisting of the land/water interface from bluff top to submerged photic zone. Juvenile Puget Sound Chinook salmon tend to favor shallow water habitat in the one- to two-meter depth range as they migrate along the Puget Sound shorelines and prepare for life in the ocean [3]. The ESA listing of Puget Sound Chinook salmon has focused significant attention on how the Puget Sound nearshore is developed, and how it can be restored to better support this culturally significant fish species. Shoreline armoring, filling, and conversion of shallow water estuary areas to uplands, and untreated stormwater discharges all have had documented negative effects on juvenile Puget Sound Chinook salmon [3]. The ESA listing has also extended renewed concern on salmon prey species such as Pacific sand lance and surf smelt that spawn on intertidal beaches. Pacific herring that spawn in eelgrass and Pacific Coast seagrass species are also important to salmon ecology [6].

Salmon are anadromous, spending part of their lives in freshwater and part in marine waters, and their survival depends upon entire watersheds, and large areas of the open ocean. As a major commercial fishery, their survival affects people in significant cultural and economic ways. This large social and ecological footprint has had major effects not only in the Puget Sound region but all of the West Coast of the United States, where a series of salmon ESA listings have occurred since 2000. The ESA salmon listings are a key factor in the funding, design, and implementation of nearshore restoration projects in the Puget Sound, and consequently a key factor in the design of Seahurst Park and Percival Landing. Meanwhile, despite significant progress being made in restoring habitat, reforming hatcheries, and implementing hydropower programs to support ESA-listed salmon populations, climate change remains a looming issue for the future of Pacific Coast salmon populations. Salmon are cold-water species that are adapted to a specific temperature range. As a possible indication of things to come, in the summer of 2015, an unusually warm, dry summer resulted in elevated river water temperatures in many parts of Washington, most notably in the Columbia River Basin, that were lethal to returning salmon, causing significant fish kills and preventing successful spawning [4].

Seahurst Park Ecosystem Restoration Project

The Seahurst Park Ecosystem Restoration Project in the City of Burien, Washington, is the largest shoreline armor removal and beach restoration project on Puget Sound, extending across nearly 1 mile of shoreline (Fig. 6.1). Initiated in 2001 by the City of Burien in partnership with the U.S. Army Corps of Engineers and many other funding partners, the park's final phase of shoreline restoration was completed in 2014 (Fig. 6.2). The ESA listing in 2000 of Puget Sound Chinook salmon and other species as "Threatened" provided a species recovery plan that was a key factor in shaping project goals and funding. The project focused on ecosystem processes and benefits for juvenile out-migrating salmon and took a comprehensive approach to planning, community/stakeholder involvement restoration and park design, monitoring and implementation. The project has a 50-year design life, and sea level rise effects were key criteria in the evaluation of the design. Located in an important area for beach sediment supply for the adjacent 10 miles of nearshore, the project utilized a habitat-forming process-based restoration approach. The park's significant existing recreation and environmental education uses were both a constraint on the potential restoration and an opportunity. Emergency access, water, sewer, and electrical utility mainlines serving off-site and on-site customers ran along and, in some cases, under the beach and shoreline.

As stated earlier, nearshore habitat in the Puget Sound region had been degraded, and, specific to the Seahurst Park landscape position, the natural bluff-to-beach sediment supply process has been significantly reduced if not completely cut off by shoreline armor. 60 % of the shorelines in the vicinity of Seahurst Park are armored [1]. Shoreline armoring has reduced the availability of intertidal beach spawning habitat for forage fish such as surf smelt and sand lance, an important food source for adult salmon. Restoring upland and shallow intertidal habitat connectivity improves the quality of beach substrates and slopes, providing additional forage fish spawning, juvenile salmon rearing, and migratory corridors [6].

Prior to the project, Seahurst Park had extensive armoring along its shoreline (Figs. 6.3 and 6.4), which greatly reduced the area's sediment supply process by cutting off the park's unstable, forested bluffs from the nearshore and the sediment transport segment (drift or littoral "cell") extending 10 miles to the north. The park's nearshore habitat was degraded from fill placement for park uses, protected by rock revetment armor, and concrete and gabion seawalls (Fig. 6.5). As a result, 3 to 4 feet of beach lowering had occurred since shoreline armor construction, and there was little connectivity between the upland riparian and intertidal habitats [2]. The project provided a rare opportunity to restore a broader ecological process and function, given the park's landscape position and the size of its shoreline.

The City of Burien's vision was significant, as they sought to plan, design, and execute a multi-objective project that integrated extensive shoreline restoration, infrastructure challenges, and significant use as a public park and environmental education facility. The City reached out to a large stakeholder group, including the local and federal project partners. The project addresses the following specific ecosystem and park goals:

Ecosystem Goals

- *Preserve existing high-functioning nearshore habitats.* Seahurst Park's northern shoreline is preserved.

- *Restore and protect the natural bluff-to-beach sediment supply process* (Figs. 6.6 and 6.7). Armoring and fill including concrete and rock gabion walls were removed, allowing for the free movement of sediment from the upland bluffs to the beach and intertidal zone.
- *Restore beach slopes and substrates.* After the removal of shoreline armoring, beach grades were restored to gradual slopes. Habitat substrate including rounded, washed gravel, and sand was placed along the beach (Fig. 6.8).
- *Restore forage fish spawning, juvenile salmon rearing, and migration intertidal habitats.* Habitat substrates were sized to accommodate fish use, particularly spawning and rearing. Driftwood was placed at appropriate beach elevations to improve habitat complexity.
- *Restore upland and shallow intertidal habitat connectivity* (Fig. 6.9). Restored backshore, wetlands, and riparian habitats provide connectivity between the aquatic terrestrial ecosystems. The planting design used multiple types of native vegetation in order to establish a diverse and multi-layered canopy (Fig. 6.10).
- *Diversify habitat in the freshwater/saltwater interface.* Freshwater input was increased through the "daylighting" of five creek mouths, restoring mini-estuary conditions that are highly biologically productive. Restored wetlands and proposed rain gardens and infiltration provide water quality treatment to stormwater prior to its discharge into Puget Sound.

Park Goals

- *Integrate shoreline habitat and recreation access.* Beach access points were strategically aligned with restoration areas to focus access in appropriate locations that protect existing and restored habitat, such as daylighted streams and coastal wetlands (Fig. 6.11). Park amenities and infrastructure, including sewer and water mains, electrical utilities, and emergency access routes, were relocated to more resilient, upland locations that are less vulnerable to sea level rise (Figs. 6.12 and 6.13).
- *Improve Americans with Disabilities Act (ADA) access and beach access.* There is ADA-compliant access throughout the site, including to the beach.
- *Protect and support educational facilities and programs.* The site's educational facilities and programs have been retained and strengthened by the project.
- *No net loss of recreational facilities.* The park's program keeps all existing park facilities, including a waterfront promenade, lawn space, picnicking area, and play area.

At first glance, the stated ecosystem and park goals could appear to conflict with one another. But this conflict and potential site constraints were overcome through creative, innovative site design and transparent, direct communication with project stakeholders and regulatory agencies. Park amenities and infrastructure, including site utilities, were relocated and consolidated to maximize wall and shoreline armor removal and beach restoration. Beach access was strategically aligned with restoration areas, to limit access to key restoration areas, such as shoreline wetlands.

Five streams were daylighted, creating multiple wetlands, and new stormwater infrastructure, such as rain gardens, was placed to improve water quality treatment (Fig. 6.14). The park's existing environmental education programs were enhanced in conjunction with the restored habitats.

The project was constructed in two phases, between 2008 and 2014. The phased approach allowed for extensive monitoring results of the smaller Phase 1 to inform the larger Phase 2. Physical monitoring showed that the Phase 1 restored beach was dynamically stable and substantiated that the beach design, including slopes and substrate, was appropriate for the wave climate, which directly influenced the Phase 2 design approach. Monitoring also informed an annual rate for the movement of beach sediment and allowed the team's coastal engineers to better develop a sustainable Phase 2 design. Furthermore, monitoring has shown that the beach sediment sizes are suitable for forage fish spawning, such as surf smelt and sand lance, which provide food for young ESA-listed salmon and are vital to the species' survival, a key ecological goal of the project.

The University of Washington and other groups conducted biological pre- and post-project scientific monitoring on Phase 1 of the project, focusing on juvenile salmon food sources such as benthic invertebrate and terrestrial insect densities and taxa richness. The research found that the park's restoration efforts have restored connections between riparian and marine habitats, and that newly converted intertidal areas have quickly become ecologically productive [5]. Consequently, Phase 2 focused on ecological gains by converting previously filled, degraded uplands to restored intertidal and riparian habitats. The monitoring has provided important lessons that extend to other nearshore habitat projects.

Seahurst Park is a destination for Burien and the greater Seattle–Tacoma region (Fig. 6.15). Two facilities on-site, the Marine Technology Lab and the Environmental Science Center, offer programs to preschool through high school-aged students on-site, for hands-on environmental and ecological education on an ongoing basis, serving 10,000–12,000 students per year (Fig. 6.16). The Seattle Aquarium also offers Citizen Science and Beach Naturalist programs at the site for the general public. And groups gather simply to picnic, access the water, and even celebrate weddings (Figs. 6.17 and 6.18). The restoration of dynamic ecosystem processes and habitats provides a strong argument that restoration and public access are compatible.

Project Partners: City of Burien (Owner), U.S. Army Corps of Engineers.
Prime Consultant: Anchor QEA.
Subconsultants: Coastal Geologic Services, Shannon & Wilson, BergerABAM, Patano + Hafermann Architects, Roth Hill Engineering Partners, Cross Engineers.
Construction Contractors Phase 1: MarVac, A-1, Segale.
Construction Contractors Phase 2: CKY.

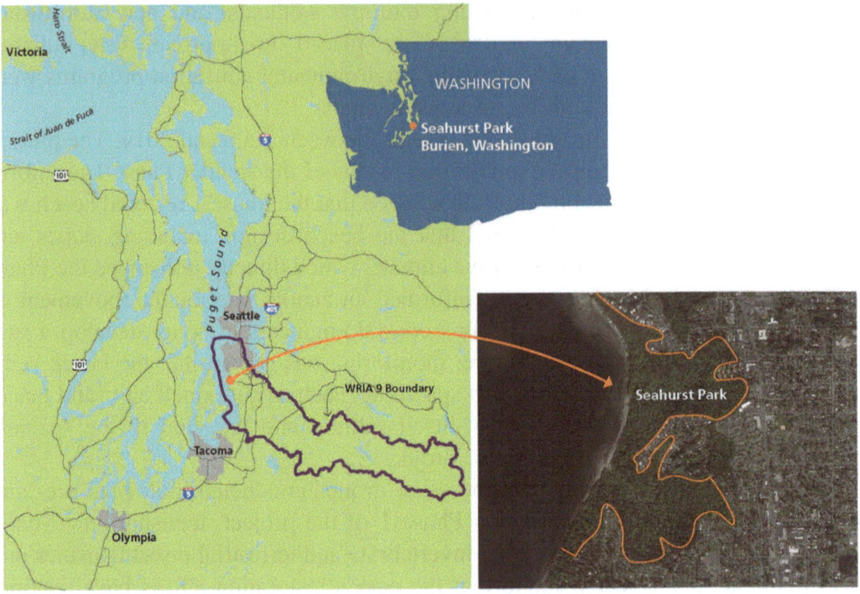

Fig. 6.1 Seahurst Park is located in Burien, Washington, a small city just south of Seattle

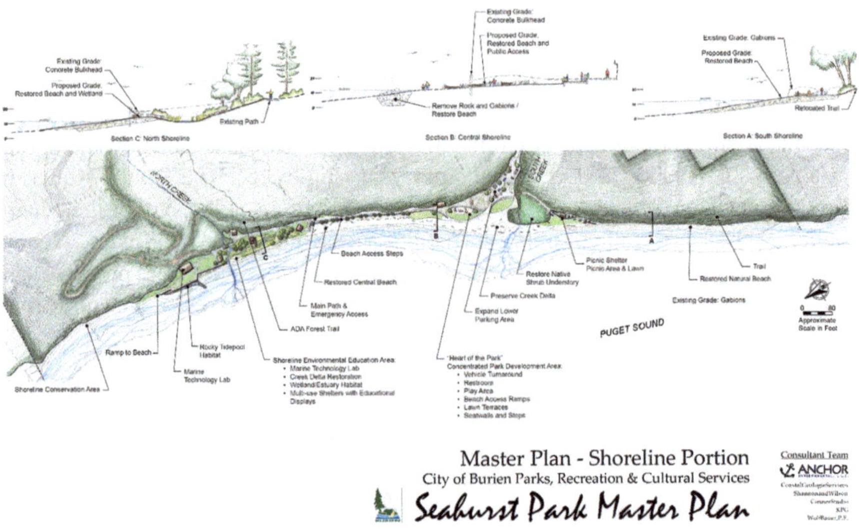

Fig. 6.2 The strategy for restoring Seahurst Park's shoreline was established in the Master Plan and implemented in phases. South Creek at Seahurst Park demarcates the boundary between the north and south shoreline phases of implementation

Fig. 6.3 View of Seahurst Park's north shoreline prior to restoration

Fig. 6.4 Prior to the project, the park's south shoreline was armored with failing rock gabion bulkheads that disconnected the shoreline from the uplands, disrupting sediment transport processes and fragmenting habitat. Similar effects occurred on north shoreline from concrete seawalls, and rock armor

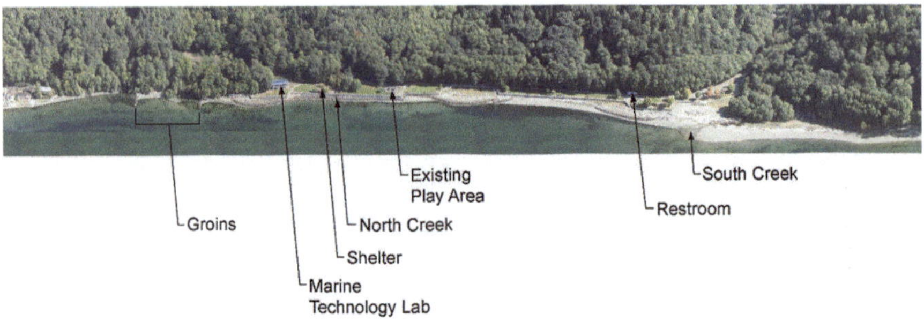

Fig. 6.5 View of existing infrastructure at Seahurst Park's north shoreline prior to the project

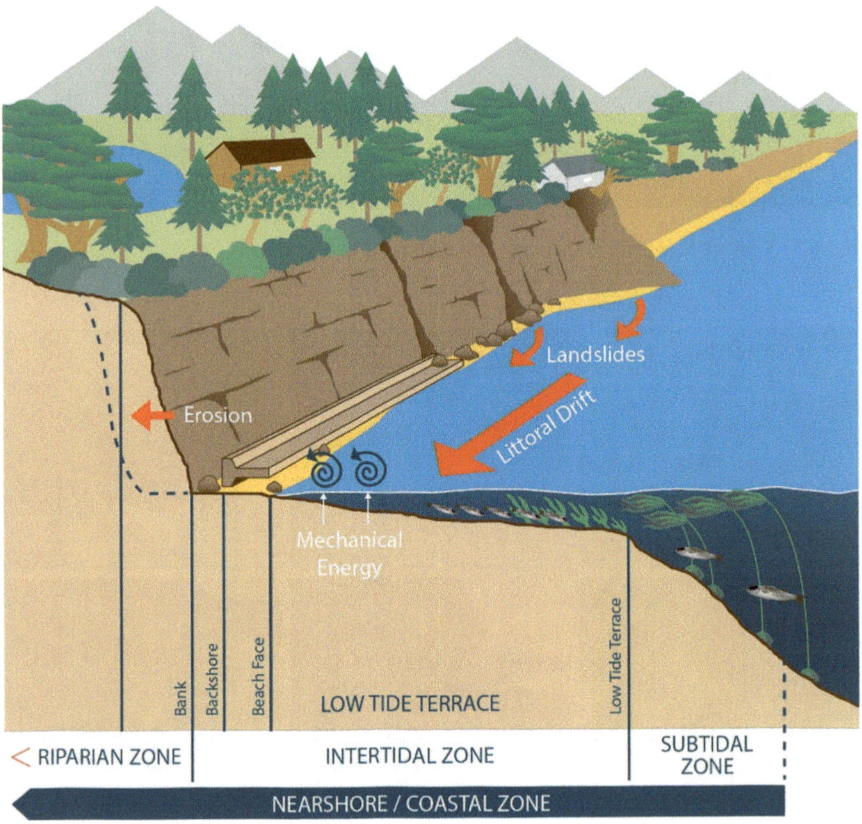

Fig. 6.6 Diagram depicting coastal sediment transport processes and bulkhead effects [2]

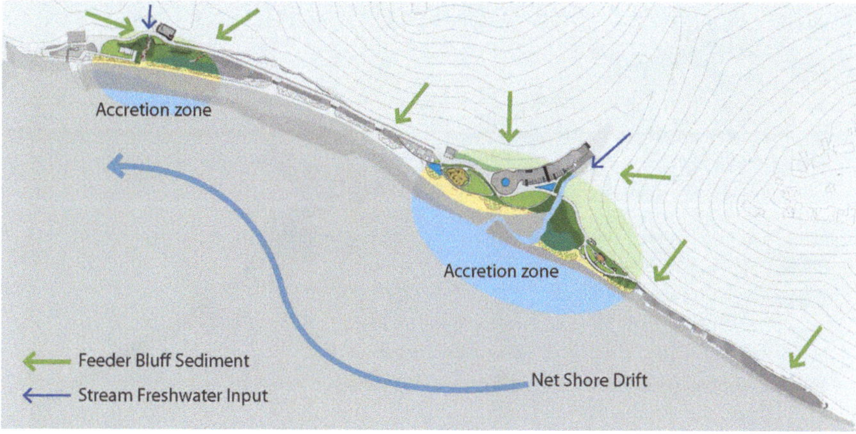

Fig. 6.7 Natural process restoration and resiliency strategy plan view

Fig. 6.8 South shoreline restoration showing gabion bulkheads that were removed. The park's shoreline was rebuilt to a more natural condition with gradual slopes, appropriate gravel and sand substrate, and reestablished riparian habitat

Other Major Funding Sources: U.S. Environmental Protection Agency, Washington State Recreation Conservation Office (Salmon Recovery Funding Board, Aquatic Lands Enhancement Account, Land and Water Conservation Fund), Puget Sound Partnership, Estuary and Salmon Restoration Program, Puget Sound Acquisition and Restoration Fund, Green/Duwamish Watershed/WRIA 9, King Conservation District.

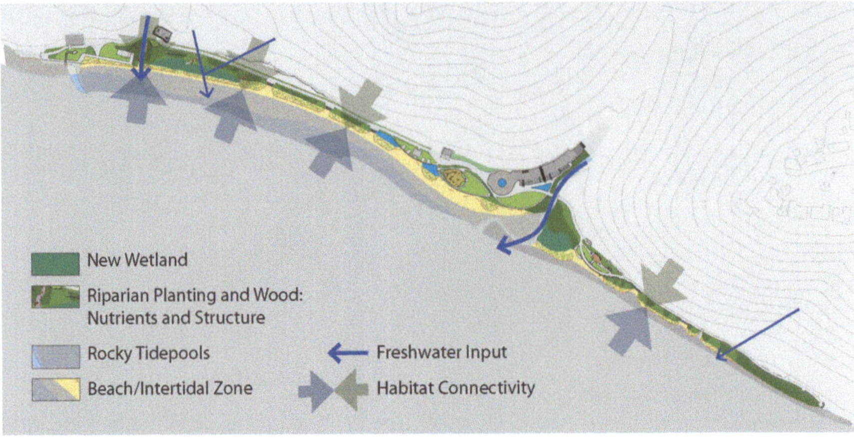

Fig. 6.9 Nearshore habitat restoration strategy plan view

Fig. 6.10 Seahurst shoreline after restoration

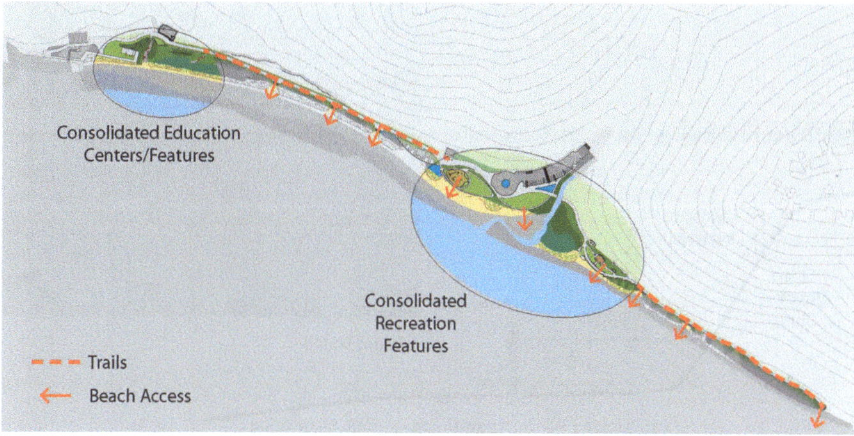

Fig. 6.11 Park amenities and public access strategy plan view

Percival Landing Major Rehabilitation Project

Percival Landing is a linear public waterfront park in the heart of downtown Olympia, Washington, the state capitol (Fig. 6.19). The site combines a nearly mile-long pedestrian shoreline access area and waterfront recreation opportunities with a commanding view corridor from the capitol to Puget Sound's Budd Inlet and the Olympic Mountains. "The Landing" is an important part of the City of Olympia's culture, economy, and environment.

The primary feature of Percival Landing was a pile-supported overwater timber boardwalk and moorage floats for transient boaters. The first portions of the boardwalk were constructed in 1978, but by 2007, most of the facilities required major rehabilitation, with some of the overwater facility closed due to deterioration and structural safety concerns (Fig. 6.20). In addition, the overwater structure and adjacent park could not sustain projected future sea level rise. Olympia is the most vulnerable city to sea level rise effects on Puget Sound, due to various factors including its low-elevation downtown built on bay fill and geologic subsidence (Fig. 6.21).

To address the vulnerability and safety issues at Percival Landing, the City of Olympia's Parks, Arts and Recreation Department undertook a multi-year process to envision the Percival Landing Major Rehabilitation Project and determine specific project goals, as defined by the citizens (Fig. 6.22). Phase 1 of the multi-phased project was completed in 2011 (Figs. 6.23 and 6.24). The following describes each project goal and how it was realized in the built work.

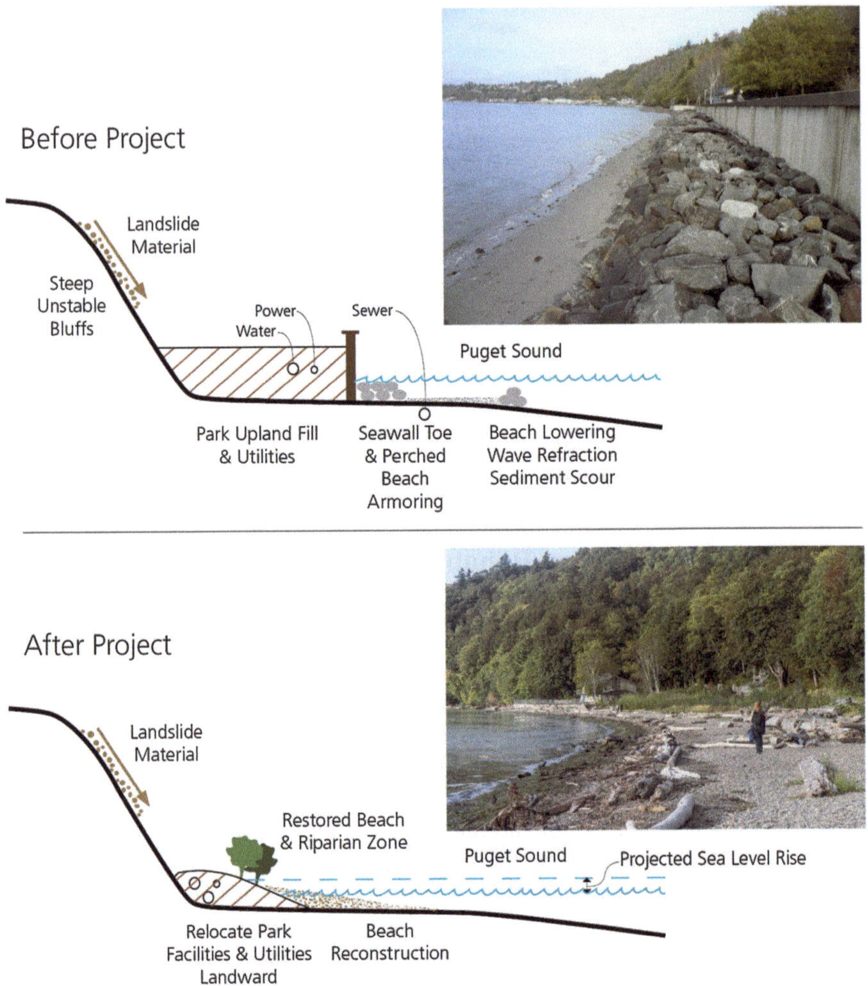

Fig. 6.12 The coastal resiliency strategy reconnected the nearshore with its adjacent feeder bluffs to restore sediment transport while still providing park infrastructure, amenities and public access

Design for a 50-Year Life, Including Predicted Sea Level Rise

Durability was a key factor in the park's design, both to withstand heavy use as a popular park, as well as withstand the harsh saltwater environment and rising seas. Concrete and steel replaced less durable materials such as treated wood.

To address the park's vulnerability to sea level rise and avoid potential inundation, landscape architects worked closely with coastal and civil engineers to analyze the threats, model future scenarios, and design the shoreline park features to

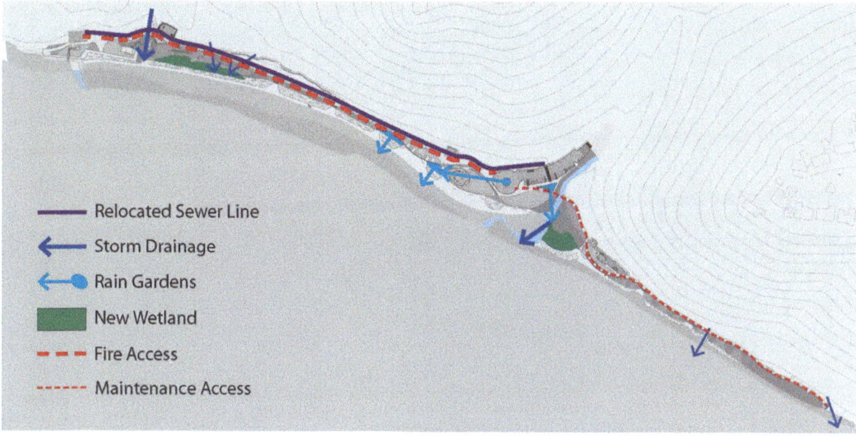

Fig. 6.13 Sustainable park and utility infrastructure strategy plan view

be resilient. The design raised finish grades, including the Harbor House finish floor elevation. Nearshore habitat grading and substrate materials were designed to withstand wind and wave energy. In addition, access to the pier utilities was modified to be from the top surface of the pier rather than from beneath.

Enhance the Natural Environment and Provide Nearshore Habitat Restoration

Puget Sound has seen dramatic degradation of its nearshore habitat in the last century, leading to the ESA listing of multiple species, including Puget Sound Chinook salmon. Nearshore ecological restoration in this urban waterfront was maximized, and the project achieved a net ecological gain in intertidal, nearshore habitat for ESA-listed salmon. Nearshore fill and debris, including broken concrete and asphalt slabs and hundreds of toxic creosote-treated timber pilings, were removed (Figs. 6.25 and 6.26). An existing waterfront parking lot was relocated upland and replaced with expanded intertidal habitat. All nearshore slopes were regraded and replaced with a fish-friendly gravel substrate or native vegetation (Fig. 6.27). All on-site stormwater is now treated before entering the Sound.

In addition, the extent of overwater structures was reduced, to lessen shading in the nearshore environment, a condition that shuts down primary production and degrades salmon habitat (Figs. 6.28 and 6.29). New overwater structures incorporated materials such as steel grating that provide more light penetration to reach the water below and improve conditions for salmon (Fig. 6.30).

Fig. 6.14 The restored North Creek now naturally supplies the beach with sediment. The creek was formerly in a culvert and sediment caused maintenance problems

Fig. 6.15 The park's resilient shoreline showing the restored North Creek

Fig. 6.16 Seahurst Park provides several environmental education programs such as the Seattle Aquarium Beach Naturalists program at low tides

Fig. 6.17 Seahurst Park offers the surrounding ethnically and economically diverse community a natural and unique gathering place

Fig. 6.18 The restored park offers the public an opportunity to interact with nature close to home

Fig. 6.19 Percival Landing is located in downtown Olympia, Washington, the state capitol

Accommodate a Wide Range of Users and Events Across a Spectrum of Ages and Physical Abilities

The rehabilitation improved the site for a full range of events, ages, and group sizes. For large events, universal pedestrian access, tent placement sites, lighting, and unobtrusive electrical hook-ups facilitate multiple types of uses, including Olympia's annual Harbor Days festival, where thousands of people gather. The large, open lawn area was improved aesthetically and functionally with expanded drainage and new perimeter seating and planting areas. The entire site was designed so that all amenities are fully accessible. Two pavilions and the Harbor House provide covered outdoor and indoor areas for school groups and families, and new outdoor seating and a play area provide spaces for people to gather and enjoy the waterfront.

Express Olympia's Unique History, Identity, and Culture

"The Landing" has a long history as a working waterfront. It was home to the Puget Sound area's first customs house for managing vessels and the trade of goods, thus

Fig. 6.20 Prior to the Project, the existing, deteriorated creosote pile-supported boardwalk was a public safety hazard and degraded nearshore habitat

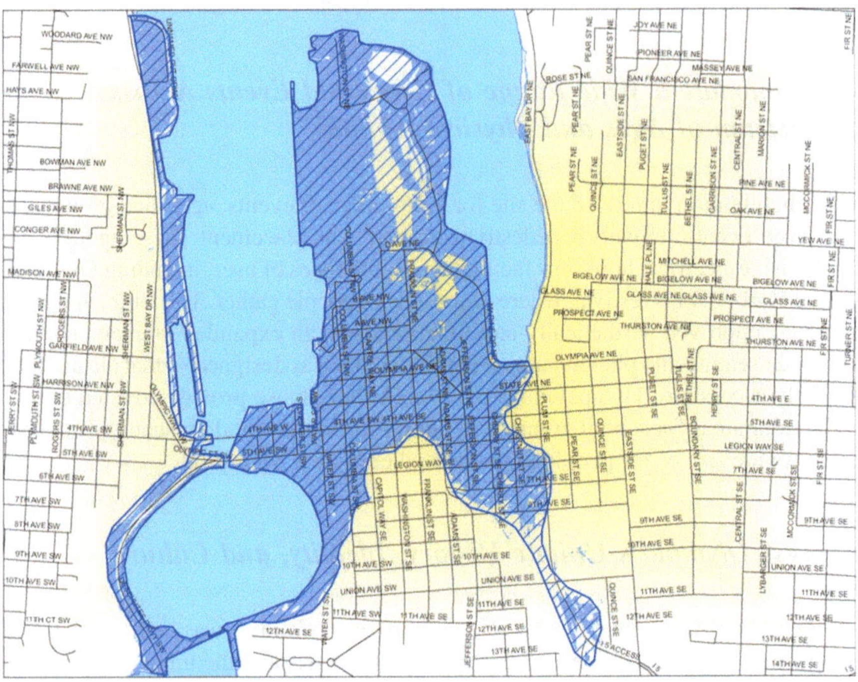

Fig. 6.21 The City of Olympia Water Resources department mapped the extent of projected tidal inundation in the year 2100 due to a forecasted 3 feet of sea level rise. Inundation is expected to effect the entire project site and much of downtown Olympia. *Source* City of Olympia, Water Resources (2007)

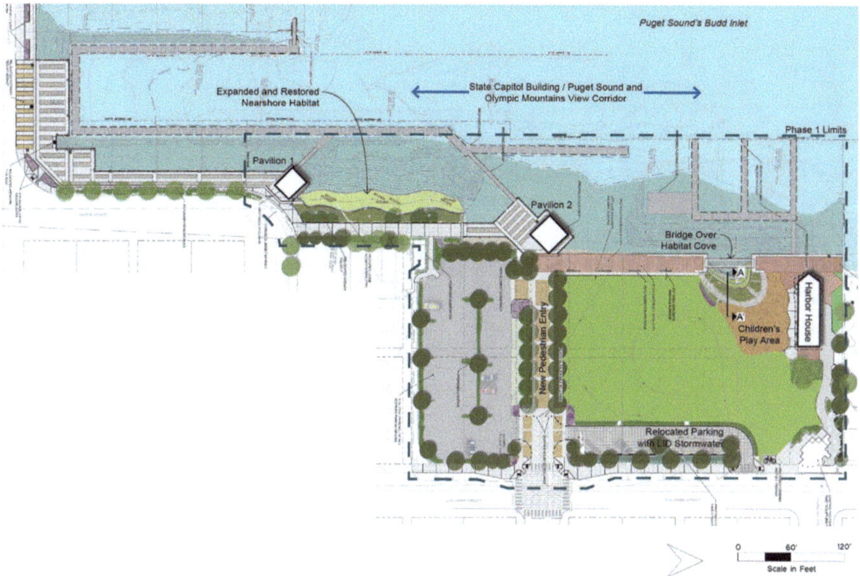

Fig. 6.22 The Percival Landing Site Plan for Phase 1 and 2 addresses the areas with the most deteriorated creosote pile-supported promenade

Fig. 6.23 The project rehabilitates the park and promenade, responds to sea level rise, enhances the shoreline's ecological function, and provides cultural amenities

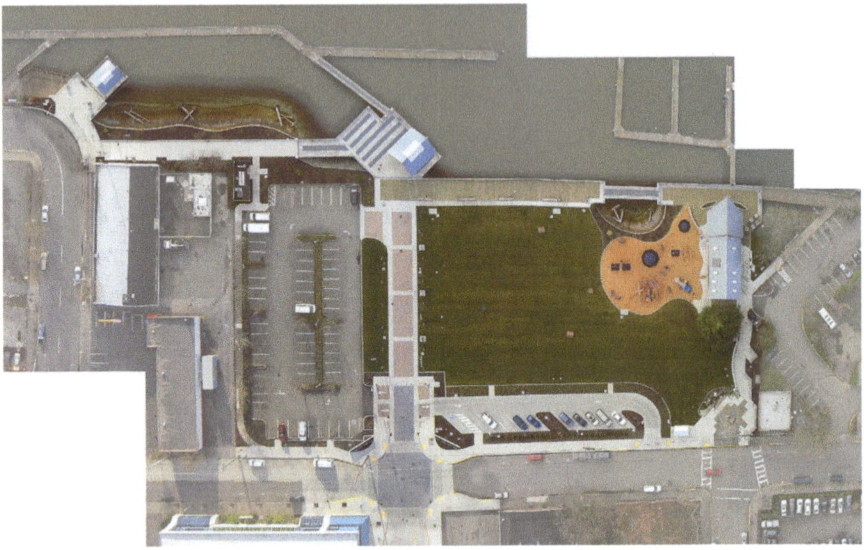

Fig. 6.24 Phase 1 rehabilitated Percival Landing aerial photo

Fig. 6.25 The waterfront park and promenade prior to the Project

Fig. 6.26 The Project site is critical habitat for federal ESA listed Puget Sound Chinook salmon. Prior to the Project, the park's shoreline habitat was degraded by creosote pilings, overwater structure shading, and concrete debris

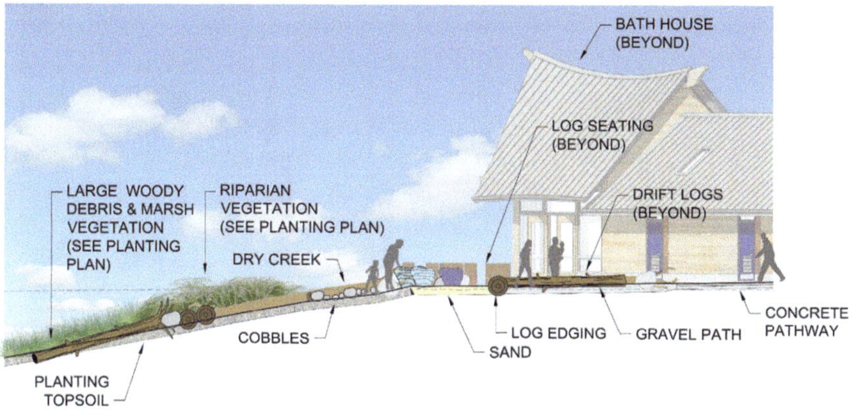

Fig. 6.27 The Project design improved ecological function of the shoreline while also providing new public park amenities

establishing it as central to the area's maritime commerce and travel. Captain Samuel Percival and, later, his son John, both leaders in the Pacific Northwest's early pioneer days, managed the commercial wharf, and the name "Percival Landing" was born. This transportation hub helped define Olympia's future role and identity as Washington's capitol.

Fig. 6.28 View north prior to the Project showing overwater creosote pile-supported structures and armoring

Fig. 6.29 View north following Project completion with promenade moved landward and restored estuarine marsh and riparian habitat

Fig. 6.30 The restored shoreline expanded estuarine marsh and riparian habitats

The project's boat hull-inspired architecture is a tribute to Olympia's working waterfront history. Two pavilions, at the intersections of street ends and the promenade, provide wayfinding landmarks for visitors with their elegant rooflines. Inside these covered spaces, interpretive and educational features are integrated around different historical themes, such as Olympia's oyster industry, and provide places to enjoy views and learn about Olympia's culture. The Harbor House provides similar way finding and educational opportunities, in addition to housing public amenities including restrooms, visiting boater showers, and a flexible meeting and event space (Fig. 6.31).

The design incorporates wood with great intention, as a means to connect users to the site's history and in locations that offer a tactile experience. The top rail of metal guard railings is wood at the pavilions and in areas where people naturally rest to enjoy the views. All benches are wood. The Harbor House and pavilion architecture highlight wood in their structure and cladding.

During the design process, citizens expressed their collective love for the site's existing sculpture, and a need for places to display sculpture or accommodate performances. In response, all existing public art pieces were preserved. Additionally, the design team worked with a public artist to locate bronze plinths, each unique, throughout the park as part of the Percival Plinth Project. Each year, the community votes on sculptures by local and regional artists for display on the plinths.

Fig. 6.31 A new cove feature provides expanded estuarine marsh habitat adjacent to the children's play area

Improve Wayfinding to and Through the Site

The former Percival Landing had multiple obstructions that confused wayfinding. In response, the landscape architect-led design team paid special attention to street ends and view corridors into the park from the surrounding neighborhood (Fig. 6.32). Character defining visual elements were placed along those corridors, and proposed new features were designed and located to avoid blocking views. For example, the two pavilions and the Harbor House have distinctive rooflines that evoke the site's history and reflect the site's unique character. These built elements are strategically sited along two street-end view corridors and along the edges of the historic capitol to the south (Fig. 6.33) and Puget Sound/Olympic Mountains view corridor to the west and north.

Provide Exemplary Environmental Performance

Through design, client advocacy, and quality construction, the project exceeded its LEED Silver target, achieving Gold certification. This was due in part to salvaging and reusing on-site materials such as the existing boardwalk timbers, a key

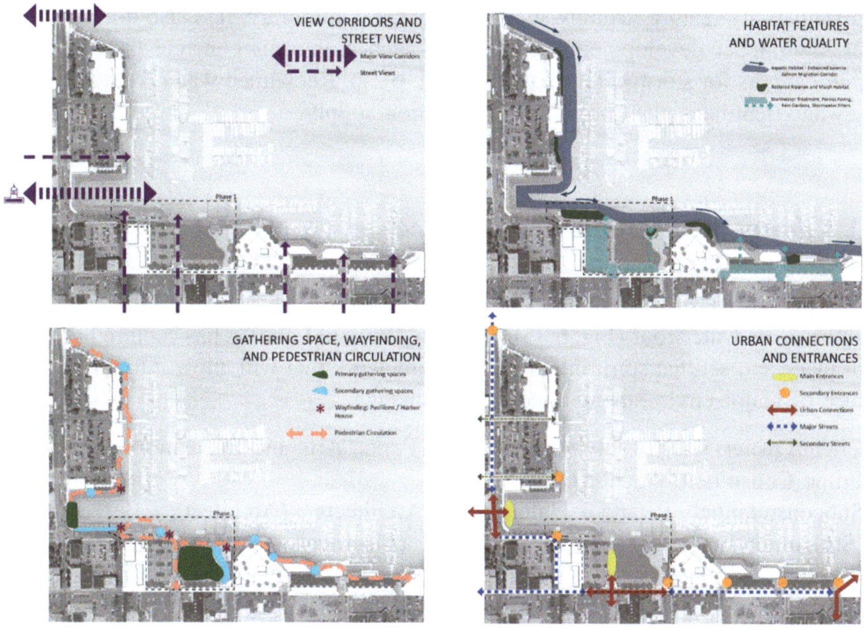

Fig. 6.32 Diagrams used during design development to understand site context and the connections and relationships between unique site elements including nearshore habitat

Fig. 6.33 The park's promenade and buildings are designed at higher elevations and built with more durable materials to address sea level rise and resiliency

sustainability feature strongly supported by the community. It is also due to the use of low-impact development technologies such as porous concrete pavement and rain gardens for stormwater treatment (Fig. 6.34). Reclaimed water from the City's sewage treatment plant provides all irrigation supply.

A Vibrant Waterfront, Rediscovered

The first phase of rehabilitation at Percival Landing has reinvigorated the City to embrace its waterfront (Fig. 6.35). The new Percival Landing has captured the spirit of place and set that spirit into a design framework that will allow it to thrive in the face of major environmental and coastal change.

Client/Owner: City of Olympia, Olympia Parks, Arts and Recreation
Prime Consultant: Anchor QEA
Subconsultants: Patano + Hafermann Architects (Architect), BergerABAM (Structural Engineer), Cross Engineers (Electrical), Northwest Archeological Associates, W&H Pacific (Survey), Greenbusch Group (Mechanical Engineers), Will Stewart Landscape (Irrigation)

Fig. 6.34 Green infrastructure stormwater solutions such as pervious paving and rain gardens helped the project achieve LEED Gold certification and improve Puget Sound water quality. This parking area was relocated landward to provide space for restored estuarine marsh and riparian habitat

Fig. 6.35 Percival Landing provides a spectacular view corridor from downtown to Puget Sound and the Olympic Mountains

Construction Contractors: Quigg Brothers, Inc. (General), Christianson, Inc., Betschart Electric, Bee Mechanical, Johnson Brothers Landscaping
Other Major Funding Sources: Washington Recreation and Conservation Office: ALEA, Washington Heritage Grant.

References

1. Anchor (Anchor Environmental) (2004) Marine Shoreline Inventory Report, Water Resource Inventory Area (WRIA) 9. Prepared for Seattle Public Utilities and WRIA 9, March
2. Anchor QEA (2011) Detailed Project Report/Environmental Assessment Seahurst Park Phase II Ecosystem Restoration Project. Prepared for U.S. Army Corps of Engineers, Seattle District, March
3. NMFS (National Marine Fisheries Service) (2007) Puget Sound Salmon Recovery Plan, Volume 1. Available from: http://www.westcoast.fisheries.noaa.gov/publications/recovery_planning/salmon_steelhead/domains/puget_sound/chinook/pugetsoundchinookrecoveryplan.pdf
4. The Seattle Times (2015) Snowpack drought has salmon dying in overheated rivers. Available from: http://www.seattletimes.com/seattle-news/environment/snowpack-drought-has-salmon-dying-in-overheated-rivers/. Published July 25
5. Toft J (2009) Benthic macroinvertebrate monitoring at Seahurst Park 2008. Year 3 post-restoration of seawall removal. Prepared for the City of Burien. SAFS-UW-90.3, July 2009

6. Washington Department of Fish & Wildlife (2015) State wildlife action plan revision—species of greatest conservation need—draft fact sheets. Available from: http://wdfw.wa.gov/conservation/cwcs/2015/draft_sgcn_fish_03–2015.pdf, March

Author Biographies

Peter Hummel, FASLA, LEED AP—Landscape Architect is a Principal Landscape Architect at Anchor QEA, a multi-discipline environmental consulting and engineering firm with over 350 scientists, planners, and engineers with offices across the United States. They specialize in water resources, surface and groundwater quality, coastal development and recreational access, habitat restoration, and contaminated sediment management projects. Mr. Hummel has over 30 years of experience focused on designing shoreline projects that serve people and benefit the environment. He thrives in an interdisciplinary setting and values the contributions of professionals in related fields. Peter Hummel's dynamic coastal landscapes function as aesthetic experiences, social spaces, and nodes of ecological enrichment. He has set a standard in the region for technically and scientifically based design, through many shoreline restoration and design projects including Percival Landing in Olympia and the Seahurst Park Ecosystem Restoration near Seattle.

He holds a BS in Landscape Architecture from the California State Polytechnic University, Pomona.

http://anchorqea.com/aqj/index.php.

http://www.nwaep.org/event-2223664.

http://www.landscapeonline.com/research/article/19842 accessed 7/11/16.

Anna Spooner—Landscape Architect is a senior landscape architect at Anchor QEA with 10 years of professional experience in landscape architecture and natural resources. Anna's background in wildlife biology and environmental planning strengthen her design approach. She has worked on multiple freshwater and marine shoreline projects that integrate public amenities into these environmentally sensitive and highly constrained natural resources. She works with multidisciplinary teams to understand the many complexities involved to realize effective solutions that consider current conditions as well as climate change and sea level rise. Through strong science and creative design, she works to balance public access with habitat restoration. Anna holds a BA in Biology from Colorado College and a MLA from the University of Oregon.

Chapter 7
Case Study: Jamaica Bay Fringing Vegetation: Restoring Upland Habitats at an Urban Shoreline

Steven N. Handel and Christina M.K. Kaunzinger

Introduction

In urban areas, past coastal change has resulted in extensive construction of infrastructure and destruction of the original vegetation. In addition, with continuing climate change, there is an influx of new plant species and plant pathogens and pests. With sea level rise and continuing climate change, the area of land available for important coastal habitats will be increasingly condensed and stressed by physical as well as biological factors. The fringing habitat adjacent to a section of Jamaica Bay, New York became available for habitat restoration when an adjacent highway was being altered. The analysis of current vegetation on this site was

S.N. Handel (✉) · C.M.K. Kaunzinger
Center for Urban Restoration Ecology, Rutgers University,
1 College Farm Rd., New Brunswick 08901, NJ, USA
e-mail: handel@aesop.rutgers.edu

C.M.K. Kaunzinger
e-mail: christina.kaunzinger@rutgers.com

M. Johnson and A. Bayley (eds.), *Coastal Change, Ocean Conservation and Resilient Communities*, DOI 10.1007/978-3-319-41914-5_7

followed by a detailed plan to install a suite of native plant species that can tolerate the predicted sea level rise and more extensive storm surges. By adding appropriate native plant species here, ecological services and wildlife habitat can be improved. The seeds of the installed plant communities can spread and enrich surrounding urban areas. Although many urban coastal areas are heavily degraded by nonnative invasive plant species, a better functioning and more biodiverse future is possible by a close collaboration between ecologists and landscape architects working with public officials. This project can be a model of how infrastructure improvement can be married to habitat improvement that is appropriate for the projected changing physical conditions.

The Landscape Problem and Potential

In many urban areas, infrastructure, residences, and commercial zones have been built right up against river, estuary, or ocean edge. Sometimes the edges are hard, a bulkhead, highway, or riprap. Sometimes a narrow soft boundary remains, often sandy, and vegetated with a mixture of botanical volunteers and installed vegetation. With sea level rise and storm surges threatening both hard and soft edges, any remaining habitat is squeezed and the already diminished ecological services are under threat. In urban zones no landforming or additional edge habitat is often available because of the constraints of occupied land inland and the great value of the marine habitat seaward. Quantity of land is constrained and will be reduced by sea level rise. Is it possible to improve the quality of the remaining habitat under these changing conditions?

The physical and biological quality of the remaining fringing habitats is impoverished. Soil conditions result from old filling operations which were concerned predominantly with raising the grade, rather than supporting biodiverse vegetation. Soil at the edge is often polluted or covered with artificial debris from storm surges and vandalism. Much of the urban flora is dominated by invasive species from abroad. Their life history characteristics allow them to sweep through native coastal vegetation, diminishing biodiversity and useful habitat for the birdlife and other animals which live at and migrate through, our urban shores. Fragments of the native coastal plant community can still be found (Fig. 7.4), but the ability of these remnants to spread is challenged by the dense stands of invasive vines, shrubs, and trees (Fig. 7.5).

Occasionally, an opportunity occurs in urban lands to improve weakened ecological structure and function. All cities have financial limitations, but new construction or other civic initiatives near the edge can allow for funds to become available to remediate past coastal habitat problems. In a world where aging infrastructure and rising tides are leading to new investment in coastal construction, there is a new opportunity to improve urban coastal habitat. The restored vegetation can reflect our new physical conditions, the need to be resilient to climate change and rising sea levels.

In addition, adding diverse native vegetation at urban edges creates a value added by displaying lovely and resilient habitats to urban populations. This can educate urban dwellers to the value of ecological structure, to the appealing natural heritage of our shores, and to improve the quality of science education at all levels. Native vegetation installed at the edge can also model what can be done in nearby home landscaping to improve ecological function and value for coastal wilds.

An opportunity to do this presented itself on the northern border of Jamaica Bay in New York City (Fig. 7.1). Reconstruction of bridges on the major road, the Shore Parkway, included funds to remediate the landscape that was damaged by this major transportation project. The controlling agencies realized that this was a chance to restore ecological health on this landscape and a partnership between restoration ecologists and landscape architects was chosen to develop a site plan to advance this goal. The City of New York Department of Parks and Recreation was the client agency. An assessment of current conditions was performed and a detailed site plan was developed to restore ecological structure and function. Work here is a model for what could be done in many other urban edges in this time of rapidly changing coastal conditions.

With relatively small parcels to work with and with defined budgets, ecological restoration projects in urban areas must be pragmatic, maximize ecological services for both wildlife and people, and garner public support so that use and management of these areas can continue in the future. The designers of this project considered both ecological and cultural value when changing a damaged and biologically depauperate landscape into one whose biodiversity and functioning could be quickly improved.

Fig. 7.1 Site location on the north shore of Jamaica Bay. Densely populated Brooklyn and a wildlife preserve surround the site

Site Ecological History

The landscape here has seen constant and extraordinary change over the past 100 years. The original irregular shoreline, lined with salt marsh grasses and shallow tidal pools was completely refigured during the development of modern Brooklyn. The shoreline was filled in with dredged sands from the area to create most of the upland upon which the Shore Parkway rests. This was a time when salt marsh and tidal zones were considered useless or dangerous to human health. Historic maps show the evolution of the landform over the past 160 years. The marshes of an 1852 map are salt meadows, with salt tolerant grasses that could persist with regular tidal flooding [16]. By the year 1900, a rail line had been built to the shore. This triggered new commercial and residential development. By 1911 the shoreline had been enlarged and straightened (Fig. 7.2). Additional landfilling along both creeks and the shore yielded a shoreline similar to today's by 1940. The current Shore Parkway was built in the 1930s.

Soil maps show this area is underlain by Big Apple fine sand and Fortress sand. These are both sandy soils generated from human transported dredged materials. Only modest vegetation was installed on this new fill. Historic design elements for the landscape adjacent to the Shore Parkway must be referenced in any plans to change the vegetation toward more resilient character [23]. Viewsheds, for example from the roadway to the bay were originally clear and must be substantially maintained (as are noted on the vegetation map of Fig. 7.7). During the next 50 years after initial construction of the roadway, innumerable woody plants invaded this landscape, some of these are species characteristic of salt shrub and maritime forest, but most are Asian species which have become abundant in urban New York. No deer, known to damage other coastal habitats [12], are present here.

In addition to the land forming and construction, the other great ecological story here is the massive growth of people in this part of South Brooklyn. The hundreds of thousands of people and their wastewater have direct and indirect impacts on the shoreline. During any day there are people wandering, fishing, or biking. Alas, many of the hiking trails and desire lines through the area are surrounded by construction and household trash and debris. Also, Jamaica Bay itself is polluted and washes trash up with every high tide and storm. The high tide line reminds one of a walk through Toys R Us, plasticware of every color and size everywhere.

These massive changes to the land stand in contrast to the still remarkable natural history of Jamaica Bay itself. Gateway National Recreation Area is the home of hundreds of bird species and millions of birds that pass through during migration seasons. The site of this project, bordering the Bay, is a place where birdlife can feed, rest, and nest. Although narrow it borders an area of high concentrations of birds because of the resources on marsh islands and protected open water in the bay. Coastal zones are now seen as valuable for ecological services beyond simply recreation [25]. The value of an intact coastal zone as protection against storm surges is now conceded as common knowledge.

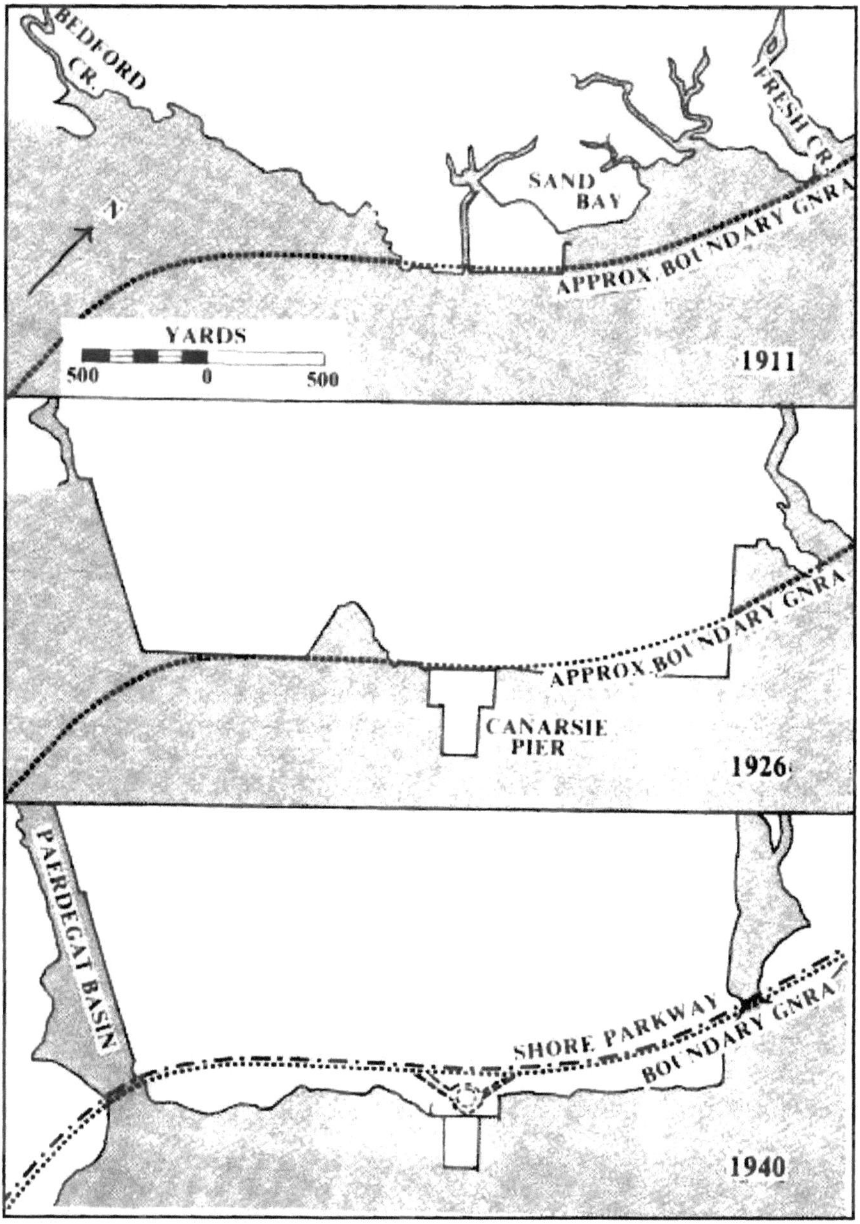

Fig. 7.2 Former and current landform (showing filling of open water and wetland). From Black [2], p. 46

Invasive species from other continents are sweeping through the East Coast, and are major elements of coastal change. Large new populations of herbaceous and woody species are the subject of massive vegetation management efforts. The urban flora of coastal New York resembles a botanic display in Asia, with many of our native species in the minority in many places [6]. The native higher plant biodiversity here is less than in natural stands which suggests that resiliency of native systems against future changes in weather and pathogens will be low. Some native plants have persisted, but in small populations.

The other dark cloud on the horizon, literally, is the rapidly changing climate. NYC's [21] climate risk information, from 2013, reports that there is a good chance that sea level will rise here 9–19 cm by the 2020s and 26–58 cm by the 2050s. The high estimate of probable change is 74 cm by the 2050s (Fig. 7.3)! The number of additional days with the maximum temperature above 90° may be 26–31 days. Currently we have 18 such days. In the winter, the number of days below freezing is predicted to be reduced by 20. Restoration of construction-damaged vegetation offers the opportunity to install species resilient to this quickly approaching climate regime.

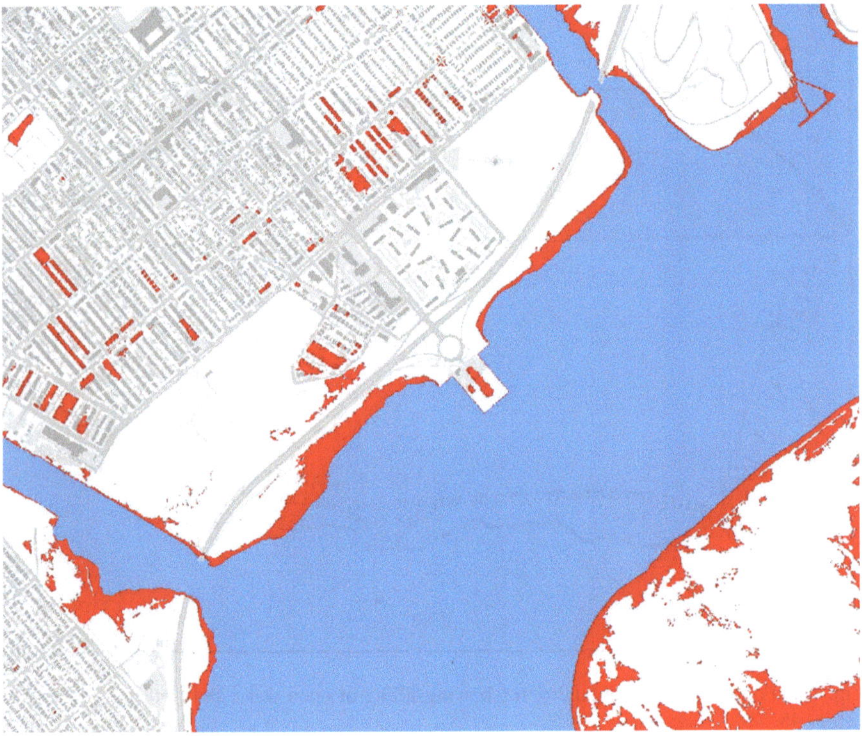

Fig. 7.3 Recent estimates of probable sea level rise over the next decades near the study site show that much of the current water's edge will be submerged, eliminating the current upland vegetation there. Storm surges will cover additional acreage. Mapping and data compilation from Piana et al. [22]

The past is not a prologue for the landscape ecology here. The predicted future will be hot, crowded, and wet. As Jamaica Bay knocks its wet fist against Brooklyn's door, all vegetation management actions must deal with a very different scenario than we have now. Government at all levels is now aware that human ecology and natural history are bound together. Actions along this stretch of the Shore Parkway must factor in significant probable change with our records of the natural history past.

Current Habitats

The site has been a fine-grained mosaic of habitat types including:

Salt Marsh; Maritime Beach; *Phragmites* stands; Salt Shrub; Floodplain Forest; Maritime Grassland; Successional Maritime Forest; Turf Grass; Bioswales. These are separated by paths and construction areas, reflecting many management activities here. Invasive species are widely spread in the habitats.

Restoration action is charged with replacing vegetation removed through construction and targeted invasive species removal near high-quality habitat and initiating a landscape that can be persistent as change in climate and sea level continues. These parcels, despite the disturbed nature of their structure, still retain wildlife habitat value. The beach and salt marsh areas could potentially support nesting by gulls, terns, American oystercatchers and black skimmers. Herons and egrets could use shrubs and trees near the salt marsh, but no signs of nesting were found. These species do not use the large patches of *Phragmites* reed. Removal of this invasive will provide room for more shrubs and trees that perching birds might use. Salt marsh expansion into the shallow open water, addition of beach sand to elevate nest sites above tidal inundation, and fencing potential nest sites to protect them from disturbance could enhance species diversity, if such actions were allowed.

The maritime forest is currently hosting only common urban and suburban bird species. Three tree species (native black cherry and nonnative tree of heaven and white mulberry) each comprise 20 % of the canopy and there are few shrubs. The understory is largely invasive mugwort and/or dense vines (oriental bittersweet, poison ivy, Virginia creeper) (Fig. 7.4). Enhancing plant diversity there would support a greater diversity of wildlife species. More native plant species would also increase the insect community to support the food web upon which most bird species depend [4, 24].

Diamondback Terrapin (*Malaclemys terrapin*) is the only North American turtle which regularly inhabits the brackish waters of coastal salt marshes, bays, tidal creeks, and ditches [3]. Once extremely common in Jamaica Bay, Diamondback Terrapins have declined due to harvest, marsh filling, and nest predation (primarily raccoon). Analysis of feasible nesting area in Gateway National Recreation Area based on availability of sun-exposed, sparsely vegetated habitat, sandy soil, and proximity to water identified this Canarsie Pier zone as potential nesting habitat [11]. However, no Diamondback Terrapin nesting has been observed [7, 11].

Fig. 7.4 Photo of current habitat invaded by many non-native plant species

The limited amount of beach above the high tide line and nest predation from raccoons are probably the limiting factors. Site enhancements could include addition of beach sand to elevate nest sites above tidal inundation, removing invasive vines from bayberry and sumac thickets to expose bare sandy soil within, and nest protection from predators. These needs are also addressed in this site plan.

The interactions among vegetation, terrestrial wildlife, and marine species adjacent to this fringing vegetation are important for overall ecosystem health [26]. Improved and resilient vegetation binds soil and absorbs nutrient flows from urban stormwater. This improves the shallow water habitats for invertebrates and marine fishes. In these ways actions on the upland fringe directly influence persistence of food web populations under the bay's water. Investment on land has value under the sea. The fish life of Jamaica Bay has commercial and recreation value, and is a focus of public interest for residents.

Framework for Ecological Restoration

The expected climate and hydrologic changes remove critical ecological niche space—microhabitats for feeding and nesting and necessary food web structure. The built-up urban adjacent space constrains inward migration of these niches. The ecological restoration of this landscape rests on a foundation that has five components:

- First, action here must result in a sustainable plant and animal community which is consistent with what we know of the past and current biodiversity of this area [1, 5, 9, 10, 15, 17, 18, 19, 20].
- Second, this landscape is heavily used by people. A major highway slices through it and thousands of people visit the site for recreation and repose. Any restoration action here must deal with human ecology, broadly defined, as well as the needs of native plants and animals.
- Third, this site has a history. This is not a natural landscape that is merely infected with modern day invasive species or an occasional storm surge which has modified some long-standing ecological structure of forest and field. In contrast, this landscape is wholly constructed and maintained by human action. The upland we see was all constructed in the 1930s. The pathways are created and used by humans, not other animals, and the large trees present were installed by landscapers. These trees were not naturally dispersed to the shore of Brooklyn. Other species (*Spartina*, bayberry, beach plum, etc.) were natural recruits in addition to those planted by the city.
- Fourth, this site is at the edge of a dynamic boundary between sea and land. Current projections are that the bay's boundary will rise, and be regularly disturbed by increased frequency of storm surges. The vegetation patterns here will be controlled by a narrow upland strand and by new saltwater intrusion regimes.
- Fifth, this parcel of land is connected ecologically to urbanized Brooklyn and to the Jamaica Bay Wildlife Refuge. The land is influenced biotically by processes and species around it. Some of these influences will be positive, such as the introduction of new species from ecologically healthy patches in the area. There are also negative influences, particularly the potential for rapid invasion by detrimental species of plants and animals (including insects) which will degrade any efforts by the land managers to improve biodiversity and sustainability. The complex and messy nature of ecological interactions is likely to result in a variety of adjustments over time and mixed compositions of native and introduced species.

These factors together require a pragmatic and resilient ecological restoration for this site. Returning to some romantic approximation of the land before European settlement is, to our eyes, a waste of time and of public funds. We considered five options as landscape restoration alternatives:

1. No action. There is a mosaic of plant communities that does in fact cover most of the land. There is ecological structure, soil is bound, modest habitat is available for birds and other animals, and a scattering of native coastal plant species is present. 20 % of the trees present, for example, are wild black cherry, an important and useful member of regional plant communities. Many patches of native coastal shrubs are found throughout the site. Leaving the landscape as is yields the no-cost alternative. However, construction of the highway and bridges has left large areas which are barren. These will be invaded by seed sources surrounding the construction zone which are predominantly alien

species of modest ecological value. There are native species on adjacent lands but restoration experience has shown that these are overwhelmed by the rapid reproductive rate of the nonnatives.

2. Surgical intervention. Ecologically, this means removing some species that are unwanted and then adding appropriate new species in the small gaps that have been created. There is a literature of the native species, trees, shrubs, wildflowers, which are appropriate for this waterfront of Brooklyn. Many of these species are available from commercial and agency sources. This approach is labor-intensive but would result in a vegetation formation which preserves the local coastal character and whose biodiversity offers greater services and functioning to the environment.
3. Remove and replace. Some patches on the site are so overwhelmed by invasive vines and aggressive trees such as *Ailanthus* (Fig. 7.5) that the most appropriate alternative would be to remove all current vegetation, remove, or kill seedlings of these troublemakers as they appear, then replace with completely new native vegetation. Different models could be used at different locations, responding to aspect and elevation above the saltwater, and we would end up with an interesting and diverse community with both resilience and function. Many studies have shown that increasing biodiversity can substantially add to site sustainability, meaning that the species will persist and would require a light hand of management. Another section of the site which could benefit from this alternative is the mowed lawn zone near Canarsie Pier. This high maintenance, low ecological value parcel could be replaced by different native formations such as meadow or shrubland.

Fig. 7.5 Photo of an intact pocket of native landscape found episodically on the site

Fig. 7.6 Photo of the roadway and beach from near one of the inlets. Construction areas surround the new roadway and bridge

4. Heal disturbed areas. The large areas of bare soil where construction activities were staged (Fig. 7.6) are available for many landscape solutions. Certainly a coastal palette, adapted to the saline soil and winds, must be chosen. However, these open areas on the right-of-way are adjacent to the lands heavily infected by aggressive alien species. Attention to only the new construction zones would give us a temporary solution soon to be overwhelmed by unwanted neighbors. Whether it is clonal growth from the reedgrass, wind carrying in *Ailanthus* seeds, or birds dispersing seed from bittersweet vine and multiflora rose, these botanical neighbors pay a visit and do not leave. Investing effort on only the construction staging areas has a low probability of yielding any ecological dividends.
5. Different strokes for different folks. At this site this old adage should be interpreted as different landscape treatments for the north and south sides of the Shore Parkway. The Parkway acts as a berm against high-energy storm surges and occasional very high tides. The north side has relative safety behind the Parkway. The south side, adjacent to Jamaica Bay, gets regular flooding. As the bay will be higher and storm surges more frequent, landscape treatment must tolerate heavy salt inundation and regular salty winds. Maritime shrub species must be the focus of any intervention here.

A final point is that action at this site can influence the quality of the vegetation *surrounding* the site. Adding native species long gone from the Canarsie neighborhood can present source material of seeds that can improve surrounding parts of Gateway National Recreation Area and in the green areas of Brooklyn. In this way action at the site can add to the resiliency and ecological quality throughout the north shore of Jamaica Bay. Investment here will conserve and enhance environmental quality more broadly. This is "value added" from the budget for this one site.

As management of the site and adjacent land is shared by state, city, and federal agencies, including actions by the Army Corps of Engineers and the Port Authority of New York and New Jersey, continuing discussions among these stakeholders will be necessary to secure long-term environmental and public health. On this particular site, the United States National Park Service controls land adjacent to the bay while the City of New York controls the land adjacent to the Shore Parkway. These two major government entities have different rules and procedures, and long-term maintenance and sustainability here will require continual communication between the officials who control this one habitat which is under the authority of two masters.

General Recommendations for Plantings South of the Shore Parkway

Our initial recommendation was for the area south of the parkway to be planted in maritime shrubland (bayberry, winged sumac, beach plum) and salt shrub (groundsel, marsh elder) as a response to increasing storm frequency and sea level rise (Figs. 7.7 and 7.8). The tree restitution nature of this project makes this option unfeasible. The agency rules state that trees destroyed (including invasive species such as *Ailanthus*) must be replaced with other trees. In its place we recommend maritime red cedar forest as a model community. Maritime red cedar forest is tolerant of salt spray and resilient in response to saltwater flooding.

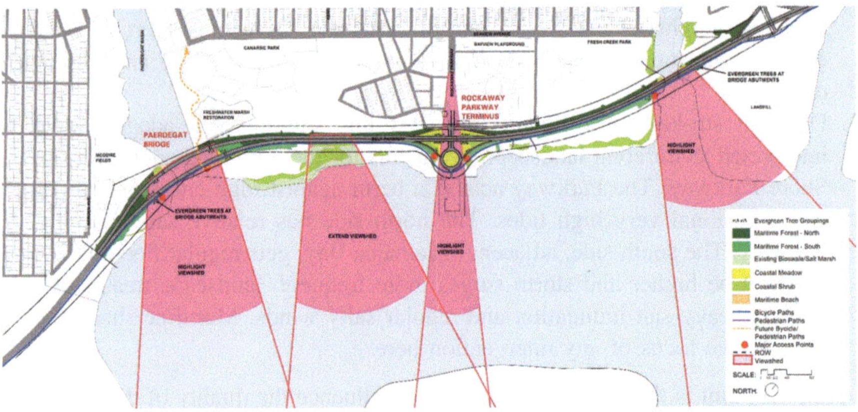

Fig. 7.7 Plan view of planting plan, from W-Architecture. A project requirement was the protection of viewshed as a cultural heritage. The color shading represents restoration areas. Green shades, from the darkest: maritime forest, north; maritime forest, south; coast shrub, bioswale/salt marsh. Yellow = coastal meadow. Brown = coastal beach

Fig. 7.8 Section of proposed vegetation, from bay to roadway then across to the northern woodland parcel

The New York Natural Heritage Program categorizes maritime red cedar forest as especially vulnerable in New York State (state rarity rank S1). Red cedar (*Juniperus virginiana*) is the dominant canopy species. At nearby Sandy Hook, NJ, black cherry (*Prunus serotina*), American holly (*Ilex opaca*) and common hackberry (*Celtis occidentalis*) are the most abundant associated canopy species. Red maple (*Acer rubrum*), blackjack oak (*Quercus marilandica*), red oak (*Quercus rubra*), and black oak (*Quercus velutina*) are also present. Common shrubs are bayberry (*Morella pensylvanica*), winged sumac (*Rhus copallinum*), and beach plum (*Prunus maritima*) with some groundsel (*Baccharis halimifolia*), marsh elder (*Iva frutescens*), pasture rose (*Rosa carolina*), and southern arrowwood (*Viburnum dentatum*). On Long Island, NY, the most abundant associated canopy species are black cherry and post oak (*Quercus stellata*). Common shrubs are bayberry and groundsel.

Coastal change affects native plantings indirectly on this site through pests and pathogens that are expanding their range as temperatures rise [8, 13]. This concern influenced the plant palette that was chosen here. For example, bacterial leaf scorch (BLS), a bacterial disease that disrupts water transport, must be considered in planting decisions for New York City. It affects urban shade trees, such as oak, sycamore, maple, and elm throughout the eastern United States; in its northern-most region (southern NY to Delaware) it especially affects members of the red oak family [14]. Planting red, blackjack, and black oak is discouraged here for this reason. Blackjack oak was included in only modest amount.

Viburnum leaf beetle, a newly introduced pest species from Eurasia on native and ornamental viburnums, is another factor to consider in planting decisions [27]. Southern arrowwood (*Viburnum dentatum*), among the most important members of the native coastal shrub community, is also among the most vulnerable species. It is included sparingly in planting recommendations for this site.

Soil pH on site ranged from 6.29 to 8.88 on existing soils, 6.4 for the construction sandy topsoil and 7.65 for the construction sandy loam topsoil. All potential species can grow at the lower end of the pH spectrum. None can grow at 8.88. Our recommendation was that soils be amended site-wide such that all appropriate species can be planted. This would contribute to a more resilient future. The secondary plan is that the construction sandy loam topsoil (pH 7.65) be amended in all locations. This pH is so high that it sets the stage for a mugwort invasion. Additionally, although some trees and shrubs may survive, seeds are unlikely to germinate and make this a self-sustaining forest. Maritime species with the lowest pH requirements have been excluded from the standard planting palettes (blackjack oak (pH: 4.6–5.6), black gum (*Nyssa sylvatica*, pH: 4.5–6), and pitch pine (*Pinus rigida*, pH: 3.5–5.1). Areas with pH above 7.5 should always have soil amendments. Some plantings require lower pH to lower the pH. They are indicated in the numbered list below.

Recommended Maritime Red Cedar Forest Palette

Trees

Juniperus virginiana, (red cedar, pH 5.0–8.0): 50 %
Celtis occidentalis (hackberry, pH: 6.0–7.8): 15 %
Ilex opaca (American holly, pH: 4.0–7.5): 15 %
Quercus stellata (post oak, pH: 4.8–7.0): 15 %
Acer rubrum (red maple, pH: 5.4–7.1): 5 %.

Understory Shrubs

Morella pensylvanica (bayberry, pH: 5.5–7.8): 30 %
Rhus copallinum (winged sumac, pH: 5.3–7.5): 30 %
Prunus maritima (beach plum, pH: 5.8–7.7): 30 %
Baccharis halimifolia (groundsel tree, pH: 5.5–7.8): 8 %
Rosa carolina (pasture rose, pH: 6.1–8.5): 2 %.

Coastal Meadow Palette

Cyperus grayi (Gray's flat sedge)
Panicum virgatum (panic grass)
Pityopsis falcata (sickle-leaved golden aster)
Prunus maritima (beach plum)
Schizachyrium scoparium (little blue stem)
Solidago sempervirens (seaside goldenrod).

General Recommendations for Tree and Shrub Plantings North of the Shore Parkway

Although this site should be protected from sea level rise, the climate change dynamics will impact the ability of local species to persist as they are stressed by hotter summers and winters that have fewer freezing days (as this effects dormancy requirements). The north side of the parkway is more protected from salt spray and overwash than the south side; it currently supports successional maritime forest (the forest patches south of the Parkway are also successional maritime forest). This forest type is an earlier seral stage of maritime post oak forest, maritime holly forest, and maritime red cedar forest. It typically exists when disturbance like wind, water, fire, or anthropogenic clearing resets the successional sequence. The land at this site is undergoing primary succession on top of the dredge material used to fill the marshes and expand the amount of buildable land into Jamaica Bay.

Maritime holly forests occur in low areas behind the dunes of barrier islands. Maritime post oak forests occur on the edges of salt marshes and on exposed bluffs and sand spits near the ocean. Maritime red cedar forest occur on dunes (as in front of the holly forest on Sandy Hook) and on the edges of salt marshes. The current successional maritime forest is far enough from the ocean that it does not have salt and wind stunted trees typical of maritime forests. The next seral stage would likely resemble a midway point between the maritime forests and the coastal oak hickory forest found on the mainland behind Sandy Hook or the variety of oak communities found on Long Island. The recommended species list draws from these communities (Figs. 7.7 and 7.8). Presence of bacterial leaf scorch disease and viburnum leaf beetle, high soil pH, and the advantages of higher biodiversity were also taken into consideration in planting recommendations.

Recommended Inland Forest Palette

Trees

Ilex opaca (American holly, pH: 4.0–7.5): 25 %
Quercus alba (white oak, pH: 4.5–6.8): 15 %
Quercus prinus (chestnut oak, pH: 4.5–7.0): 15 %
Quercus stellata (post oak, pH: 4.8–7.0): 15 %
Amelanchier canadensis (serviceberry, 5.6–7.5): 10 %
Sassafras albidum (sassafras, pH: 4.5–7.2): 10 %
Juniperus virginiana, (red cedar, pH 5.0–8.0): 2.5 %
Acer rubrum (red maple, pH: 4.7–7.3): 2.5 %
Celtis occidentalis (hackberry, pH: 6.0–7.8): 2.5 %
Rhus typhina (staghorn sumac, pH: 4.5–7.2): 2.5 %.

Understory Shrubs

Gaylussacia baccata (black huckleberry, pH: 4.5–6.5): 28 %
Morella pensylvanica (bayberry, pH: 5.5–7.8): 28 %
Vaccinium corymbosum (highbush blueberry, pH: 4.0–6.5): 28 %

Photinia pyrifolia (red chokeberry, pH: 5.1–6.5): 5 %
Viburnum dentatum (arrowwood, pH: 5.1–6.5): 4 %
*Sambucus canadensis (*elderberry, pH: 6.1–7.5): 3 %
Rhus copallinum (winged sumac, pH: 5.3–7.5): 2 %
Rhus glabra (smooth sumac, pH: 5.3–7.5): 2 %.

Together, these are plants which are found in coastal habitats in areas to the south, as well as in New York City. Consequently, planting stock could be found from somewhat southerly landscapes that now have the temperature regime that New York City will experience in the decades ahead. These species are "pre-adapted," to the future local climate by having populations whose provenance and range includes warmer areas to the south. Rather than looking west for future success, as Horace Greeley once wrote, restorationists should be looking south, for plant species and planting stock that can tolerate the expected warmer conditions of the coast in the rest of this century.

Acknowledgments We thank the staff of W-Architecture in New York City, Barbara Wilks, Julia Howe, and Martin Barry for leading the design work of this project. Elizabeth Jordan and Mary Zervos of the NYC Department of Parks and Recreation steered us through the regulatory and municipal requirements for this site. The Rutgers field staff spent long hours assessing the site and developing the botanical character of the landscape in great details: Lea R. Johnson, Kevin P. Corrigan, Thomas Young, and Josh Echols. Dr. Stephanie Murphy of the Rutgers Soil Testing Lab offered valuable help with understanding the soil needs here.

References

1. Bertness MD (2007) Atlantic shorelines: natural history and ecology. Princeton University Press, Princeton
2. Black FR (1981) Jamaica bay: a history. Study No. 3. Division of Cultural Resources, North Atlantic Regional Office, National Park Service, U.S. Department of the Interior, Washington, D.C.
3. Brennessel B (2006) Diamonds in the marsh: a natural history of the diamondback terrapin. University Press of New England, Hanover
4. Burghardt KT, Tallamy DW, Gregory Shriver W (2009) Impact of native plants on bird and butterfly biodiversity in suburban landscapes. Conserv Biol 23(1):219–224
5. Chrysler MA (1930) The origin and development of the vegetation of Sandy Hook. Bull Torrey Bot Club 57:163–176
6. Clemants SE, Moore G (2005) The changing flora of the New York metropolitan region. Urb Habitats 3(1):192–210
7. Cook R (1989) A natural history of the Diamondback Terrapin. Underw Nat 18:25–31
8. Dukes JS, Pontius J, Orwig D, Garnas JR, Rodgers VL, Brazee N et al (2009) Responses of insect pests, pathogens, and invasive plant species to climate change in the forests of northeastern North America: What can we predict? Can J For Res 39(2):231–248
9. Edinger GJ, Evans DJ, Gebauer S, Howard TG, Hunt DM, Olivero AM (eds) (2002) Ecological communities of New York State, 2nd edn. New York Natural Heritage Program, New York State Department of Environmental Conservation, Albany, NY
10. Edinger GJ, Feldmann AL, Howard TG, Schmid JJ, Eastman E, Largay E, Sneddon LA (2008) Vegetation classification and mapping at Gateway National Recreation Area.

Technical Report NPS/NER/NRTR—2008/107. National Park Service, Northeast Region. Philadelphia, PA
11. Feinberg JA, Burke RL (2003) Nesting ecology and predation of Diamondback terrapins, *Malaclemys terrapin*, at Gateway National Recreation Area, New York. J Herpetol 37:517–526
12. Forrester JA, Leopold DJ, Underwood HB (2008) The role of disturbance in the long-term viability of a coastal maritime forest fragment. Technical Report NPS/NER/NRTR–2008/104. National Park Service, Boston, MA
13. Garrett KA, Dendy SP, Frank EE, Rouse MN, Travers SE (2006) Climate change effects on plant disease: genomes to ecosystems. Ann Rev Phytopathol 44:489–509
14. Gould AB, Hamilton G, Vodak M, Grabosky J, Lashomb J (2004) Bacterial leaf scorch of oak in New Jersey: Incidence and economic impact. Phytopathology 94:S36
15. Harshberger JW (1900) An ecological study of the New Jersey strand flora. Proc Acad Nat Sci Phila 52:623–671
16. Hendrick DM (2006) Images of America, Jamaica Bay. Arcadia Publishing, Charleston
17. Kieran J (1959) A natural history of New York City. Houghton Mifflin Company, Boston
18. Kiviat E, Johnson EA (2013) Biodiversity assessment handbook for New York City. American Museum of Natural History, Center for biodiversity and Conservation, New York, and Hudsonia Ltd., Annandale
19. Linnaean Society of New York (2007) Natural history of New York City's parks and Great Gull Island. Linnaean Society, New York
20. Martin WE (1959) The vegetation of Island Beach State Park, New Jersey. Ecol Monogr 29:1–46
21. New York City Panel on Climate Change (2013) Climate risk information 2013: observations, climate change projections, and maps. In: Rosenzweig C, Solecki W (eds) NPCC2. Prepared for use by the City of New York Special Initiative on Rebuilding and Resiliency, New York
22. Piana M, Meixler M, Kleinbeck A (2016) Modeling the projected impacts of sea level rise on ecosystem services of the urban-coastal fringe in Jamaica Bay, New York. (In Preparation)
23. Marion P (2012) Historic inventory report. Study for NYC Department of Parks and Recreation
24. Tallamy DW (2009) Bringing nature home: how you can sustain wildlife with native plants. Timber Press, Portland
25. TEEB—The Economics of Ecosystems and Biodiversity (2011) TEEB manual for cities: ecosystem services in urban management. www.teebweb.org
26. United States Fish and Wildlife Service (1997) Significant habitats and habitat complexes of the New York Bight watershed: Jamaica Bay and Breezy Point. Complex#16. Charlestown, RI. http://nctc.fws.gov/resources/knowledge-resources/pubs5/web_link/text/toc.htm#16
27. Weston PA, Desurmont G, Hoebeke ER (2007) Viburnum leaf beetle. Am Entomol 53(2)

Author Biographies

Steven N. Handel—Restoration Ecologist is Distinguished Professor of Ecology at Rutgers University and Visiting Professor of Landscape Architecture at Harvard University. He is a restoration ecologist studying habitats and biodiversity in urban and coastal areas. His background is in plant population ecology and plant–animal interactions. He has collaborated with landscape architects on the design of urban parks. He serves as Editor of the journal Ecological Restoration. Honors include being an Aldo Leopold Leadership Fellow of the Ecological Society of America and a Fellow of the American Association for the Advancement of Science and Honorary Membership to the American Society of Landscape Architects for "nationally or internationally significant achievements." In 2011, he received the Theodore Sperry Award by the Society for

Ecological Restoration for "pioneering work in the restoration of urban areas." He received his B.A. in biology from Columbia University and Ph.D. in the Field of Ecology and Evolution from Cornell University.

http://deenr.rutgers.edu/facultylist.html

http://larchwp.be.washington.edu/uwla-departmental-lecture-steven-handel-rutgers-university/

Christina M.K. Kaunzinger—Restoration Ecologist is a community restoration ecologist, with advanced training in species interactions. She works on coastal habitats, inland woodlands, and environmental education initiatives. Recent projects include the shoreline along Jamaica Bay and the restoration of the Fernbank Museum forest in Atlanta, Georgia. Also, she headed a public communications effort for the Duke Farms Foundation in NJ. She was the lead ecologist for the "Dryline" plan to protect the shores of Manhattan from future sea level rise. This plan received the ASLA Honor Award in Analysis and Planning, the APA Design Excellence Award, and the AIA Honor Award for Regional and Urban Design. This project meshes ecological, protective, and social improvements. She has taught university-level courses in plant, marine, and restoration ecology, vertebrate zoology, and limnology at Rutgers University and Drew University. She received a Ph.D. in ecology and evolution from Rutgers University and a B.A. in biology from Drew University.

http://www.srijb.org/sotb2016/biographies/

Chapter 8
Case Study: Establishing Ecological Trajectory and Design Solutions for the Ash Creek Estuary

Bryan Quinn

Ash Creek Estuary

B. Quinn (✉)
One Nature LLC, 178 Main Street, Beacon, NY 12508, USA
e-mail: bryan@onenaturellc.com

M. Johnson and A. Bayley (eds.), *Coastal Change, Ocean Conservation and Resilient Communities*, DOI 10.1007/978-3-319-41914-5_8

The Ash Creek tidal estuary is one of the few remaining tidal estuaries in southwest Connecticut that is relatively undeveloped. Located in Bridgeport and Fairfield, Connecticut, it provides a tremendous number of important ecological functions, serving as a sanctuary for nesting birds, shellfish, and finfish. It also contains a large amount of unditched *Spartina* marshland. In addition to its wildlife and plant habitat, the estuary provides many other services to the community. It provides surrounding areas with flood control, shoreline stabilization from its tidal wetland vegetation, an opportunity for recreation, and an aesthetic identity that contributes to higher property values, and ultimately a higher quality of life.

Recent History

In the 1960s, a portion of the Ash Creek estuary was dredged to construct a new marina. A large jetty was constructed around the same time to protect boats as they traveled through the narrow mouth of the estuary to a newly constructed marina. This construction significantly altered the small, protective barrier spit which had historically been a sandy landform with shallow water in which it is easy to wade. The new jetty, combined with later dredging in the navigable channel, created dangerous tidal flows. It also altered the way sand migrated up and down the coast, essentially starving the landform of its essential land-forming elements. As a result, the beach became much coarser and rocky.

Following the Clean Water Act (1967) and the Tidal Wetlands Act (1969), development in the Ash Creek Estuary slowed significantly and its natural resources received more attention and protection. But in 2003, when a series of very large docks were proposed to brig motorized vessels into the shallow estuary water, a new small group of citizens joined together in opposition. They called themselves the Ash Creek Conservation Association (ACCA), and they have successfully reduced planned dock size, sponsored habitat restoration work and prevented the installation of an unsightly, and potentially environmentally damaging, utility bridge. The Association is ideally situated, and uniquely qualified, to be a link between the municipalities of Bridgeport and Fairfield, ultimately playing a central role in the development and coordination of planning efforts for the estuary.

Ash Creek Master Plan Identifies St.-Mary's-by-the-Sea Improvement Project

In 2012, One Nature created a Master Plan for the Ash Creek Estuary as a whole (see Fig. 8.1), as well as an intensive analysis of the site's eco-history [1]. The work conducted led to the establishment of a community wish list of projects. Of these

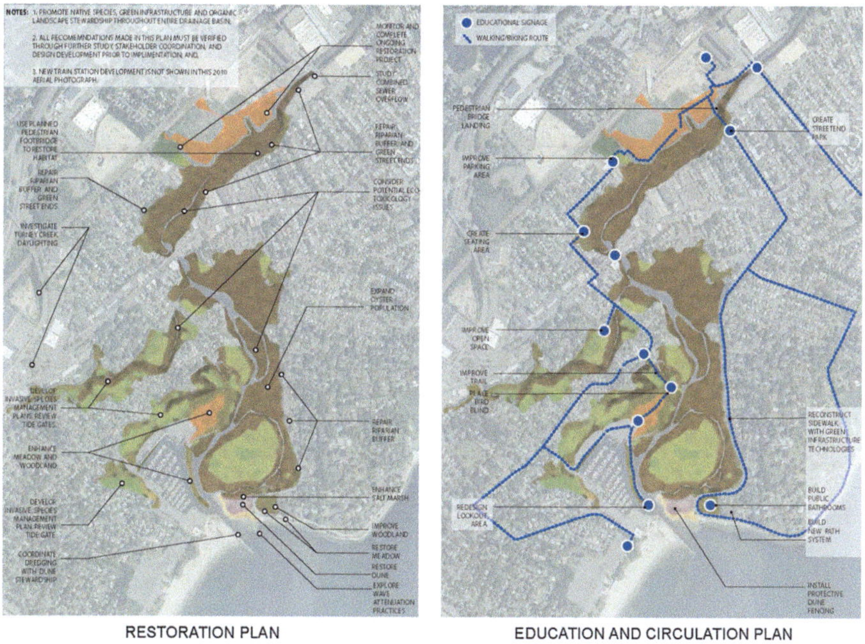

Fig. 8.1 The Ash Creek Estuary Restoration plan and education and circulation plan visually locate the various community wish list projects that were established during the estuary master planning process

projects, improvement of the St.-Mary's-by-the-Sea barrier spit, was selected by community partners to be the highest priority due to the imminent threat from coastal erosion. St. Mary's sand spit and St.-Mary's-by-the-Sea shorefront are located within the tidal inlet of Ash Creek, located at the confluence of the Creek with Long Island Sound.

Site Analysis and Establishing Ecological Trajectory

After completing an initial natural resource inventory, One Nature worked to establish an ecological history of the estuary. Over a series of months, we established a historical timeline for Ash Creek from the last glacial epoch [2].

Timeline of Historic Events Impacting Ash Creek

DATE	EVENT	IMPACT
20,000 BCE to 8,221 BCE	Glacial advance	Modification of landforms through glacial scouring and decreased seal levels.
8,220 BCE to 3100 BCE	Glacial retreat	Deposition of geologic materials, plant colinization, and rapid sea level rise.
3099 BCE to 1638	First human impacts	Native american civilizations eliminated most terrestrial megafauna and modified the landscape through agricultural activities.
1639 to 1859	European settlement	Replacement of native american communities with more resource intensive developments. Modification of shoreline and coasts to enable enhanced ocean-based industry.
1860 to 1939	Industrial Revolution	Watersheds bisected by rail lines, water and air pollution, toxic manufacturing chemicals introduced into the environment, potent levels of human sewage and waste drain to the Ash Creek estuary.
1939 to 1979	Car culture establishment	Marina construction, extensive systems of roads (esp. I-95) and new residential developments are built throughout the watershed, most streams culverted or buried, pesticide and herbicide use becomes common.
1980 to Present	Environmental advocacy and regulation	Various new and existing laws and policies are enforced at the local level to prevent further dredging, docks, and development. Formation of the Ash Creek Conservation Alliance. Completion of the Estuary Master Plan.

This analysis generally showed a gradual impairment of ecological function beginning with the industrial revolution [3, 4]. Major impacts included

- The alteration of surface water movement by coastal railroad line, the Post Road, and Interstate 95;
- Upstream pollution and increased storm water runoff due to residential and industrial development; and,
- The introduction of invasive species, especially *Phragmites australis*, Common Reed [5].

In order to conduct a thorough analysis of the benefits and threats associated with St.-Mary's-by-the-Sea, and the Ash Creek Estuary, One Nature conducted an oral history about the area by recording interviews with various community members about how the barrier spit has changed over time. This valuable input has been able to inform recommendations for restoration and protection efforts.

St. Mary's Barrier Spit and Shoreline

The St. Mary's sand spit is a type of coastal barrier landform known as a barrier spit. Barrier spits are coastal barriers that are attached to the mainland at only one end and extend into open water. Barrier spits can become barrier islands if they detach completely from the mainland, and conversely, a barrier spit can attach on both sides to the mainland and be called a bay barrier. The St. Mary's sand spit occupies approximately 2.5 acres at high tide. The sand spit faces both inland toward the tidal creek and outward toward the Long Island Sound. The Sound-facing portion of the sand spit is a fore (sand) dune environment, where vegetation is primarily characterized by low growing herbaceous plants, shrubs, and a few trees. These plants are rooted in deposits of unconsolidated loamy and coarse sand. The physical shape of taller species is impacted by the predominant onshore winds. Between the dune vegetation and the Sound, is an intertidal zone of vegetated beds.

St.-Mary's-by-the-Sea currently consists of approximately 660 ft of predominantly rocky intertidal shorefront, though the furthest western portion of this area is sandy. Above the slope to the shorefront there is a promenade with a walking path, and a grassy strip with park benches. Grovers Avenue runs parallel to this promenade with parallel parking spaces on the Sound side of the street. The promenade is part of the City of Bridgeport St.-Mary's-by-the-Sea Park. Benches, streetlights, and occasional trash receptacles can be found up and down the promenade (Figs. 8.2, 8.3 and 8.4).

Fig. 8.2 A panoramic view of the St.-Mary's-by-the-Sea Barrier Spit from the western project boundary

Fig. 8.3 Saint Mary's Chapel, the barrier spit's namesake, is shown here in 1888. Note the more gradual dune profile and finer sand material which have been lost to erosion in the past century

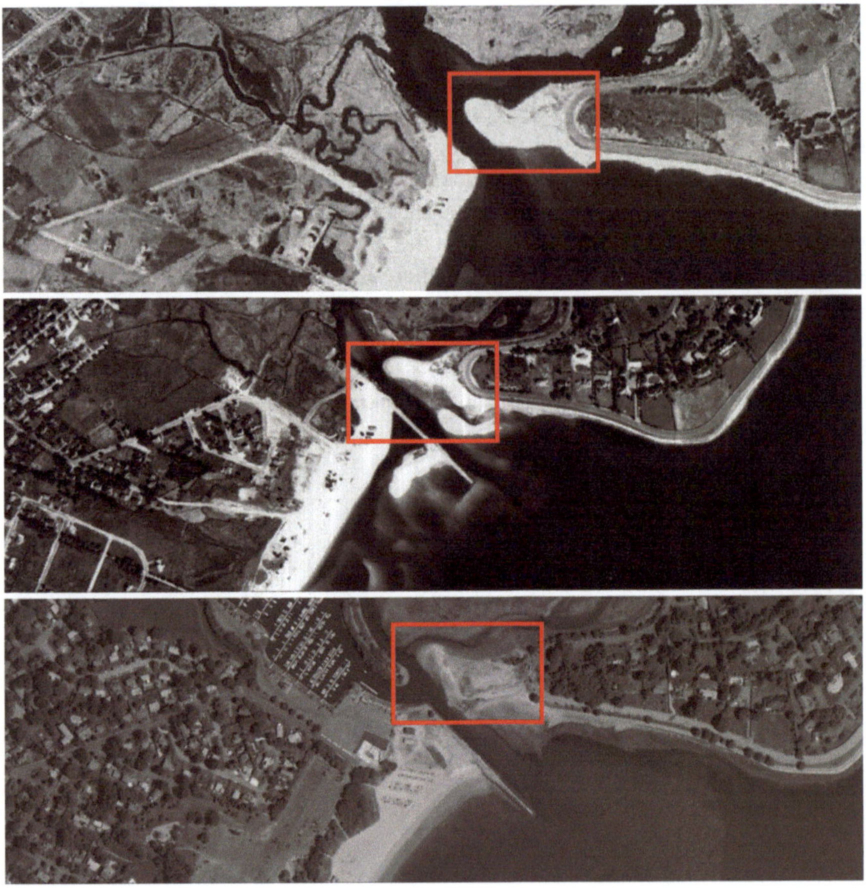

Fig. 8.4 From top to bottom, aerials of St. Mary's spit in 1934, 1952, and present-day. Note the addition of a jetty and marina in the 1952 aerial. *Red box* indicates location of proposed improvement project

Beneficial Role of St. Mary's

The benefits that the St. Mary's sand spit provides affect not only the St.-Mary's-by-the-Sea Park, but also the entire Ash Creek Estuary, and thus, the communities nearby. The location of the estuary is along the Atlantic Flyway, making it a prime stopover and feeding location for migratory shorebirds along the Connecticut shoreline. The tidal estuary is also an important habitat for seed oyster and hard shell clam beds.

The sand spit plays an important role in the function of the creek, notably by providing sand and dune habitat and by protecting the inner creek from erosive wave action during storm events, therefore mitigating flooding to the inner creek and to inland structures. Without the shelter from the sand spit, the estuary's Great Salt Marsh Island would erode away. The sand spit also serves as a sand storage area that supplies sand to eroded beaches during storms, and serves as a buffer to block windblown sand and salt spray from the inner shores. The sand dunes located in the middle of the spit are an important component of the sand spit, and are absolutely vital to the protective function of the spit. By being at a higher elevation than the rest of the sand spit, they absorb the impact of storm surges and high waves. The dunes are created and maintained by windblown sand that becomes trapped by the vegetation. Over time, the sand accretes into dunes.

The sand spit provides protection for the area, by buffering wave action during storms. Recreationally, the area provides walkers, bikers, dog walkers, and fisherpeople with a beautiful public space in close proximity to an urban area. The seaward views, the promenade, and availability of parking makes the St.-Mary's-by-the-Sea an attractive resource for many. The barrier spit serves as a primary anchor point for the surrounding culture of Black Rock, a neighborhood of Bridgeport. St. Mary's spit serves as a beautiful landform in the spirit of Long Island sound dune systems, and has become a place of community pride.

Potential Threats to St. Mary's Sand Spit

One of the greatest threats to the sand spit is erosion. The underlying sand spit deposits are subject to erosion during severe storms. Erosion has been accelerated in recent years due to human-related activities which have limited littoral drift of sand. Recent large storm events (such as Tropical Storms Irene, Lee and Sandy) have all negatively impacted the width of the landform. The long-term stability and therefore the health barrier spits have historically depended on the replenishment by sand carried by the offshore currents moving westerly along the edge of the coast. However, this normal process of littoral drift has been interrupted due to diversions from groins, jetties, or other hardened structures. As a result, there is very little sand transport, and the sand spit has degraded over time. Both of these factors are of concern to the future of this landform. Upcurrent and downcurrent coastal features, such as the St.-Mary's-by-the-Sea Walkway and Jenning's Beach, also show signs of erosion damage.

Any activity such as trampling or unnecessary development, which disturbs the natural vegetation of the sand spit, will also ultimately harm the sand dunes, since the mat of native grasses required to create and maintain the sand dunes is easily damaged by human foot traffic.

Another threat to the area is the expansion of populations of alien and non-native plant species. Non-native species have been observed in the barrier spit area. Alien and invasive species pose a threat to certain types of native plants and animals, such as the native dune grass, *Ammophila breviligulata*, American Beachgrass.

St. Mary's Concept Plan

Following extensive ecological, cultural, and historical analyses, it was determined that the St.-Marys-by-the-Sea barrier spit could be partly restored and partly rebuilt into a more resilient landform. One Nature presented its findings to the Black Rock Community and, with the support of ACCA, were awarded funding to develop a restoration plan for the barrier spit that integrated both human and nonhuman requirements.

Generally speaking, it is not economically feasible to replenish all of the threatened beaches of the region, so it is necessary to choose target restoration areas wisely. St.-Mary's-by-the-Sea meets common sense economic criteria for sustainable coastal reconstruction because it would require a relatively small amount of sand that would have a relatively big impact on the estuary's ecological functions.

One Nature's proposed concept plan for the barrier spit features a restored salt marsh, dune fencing around a raised dune with plantings for coastal meadow and beach grass, an improved woodland aside the coastal meadow, and a planted buffer next to the restored marsh. Overall, the existing high point of the spit would be lifted approximately four feet (Fig. 8.5).

Once complete, the project would create a tremendous aesthetic improvement to the community. Native grasses adapted to salt spray, storms, and coastal dynamics would lushly cover and help to build the dune; visitors would have the opportunity to walk the high tide line around the barrier spit year round. Improvements to adjacent woodland, a new picnic area, and better linkages to the St. Mary's walkway would further enhance the beauty and use of the space.

The project's budget (see Fig. 8.6) is comprised of the materials needed for reconstruction and restoration. These materials include dredged sand, plants, and dune fencing. Our estimate's largest costs included 3900 cubic yards of beach sand and 42,500 native plants. These materials, coupled with the costs of mobilization and other sitework, amounted to a $255,990 total estimate.

The cost of sand—the largest project cost—could be reduced through the coordination of ongoing dredging activities by a neighboring marina. The adjacent marina currently dredges a channel between the barrier spit and the marina to maintain a navigable channel. This sand is typically reused at the town of Fairfield's Jennings Beach Park. However, it would be more economical to directly place the material on the barrier spit. Such an agreement would reduce overall project costs up to 50 %.

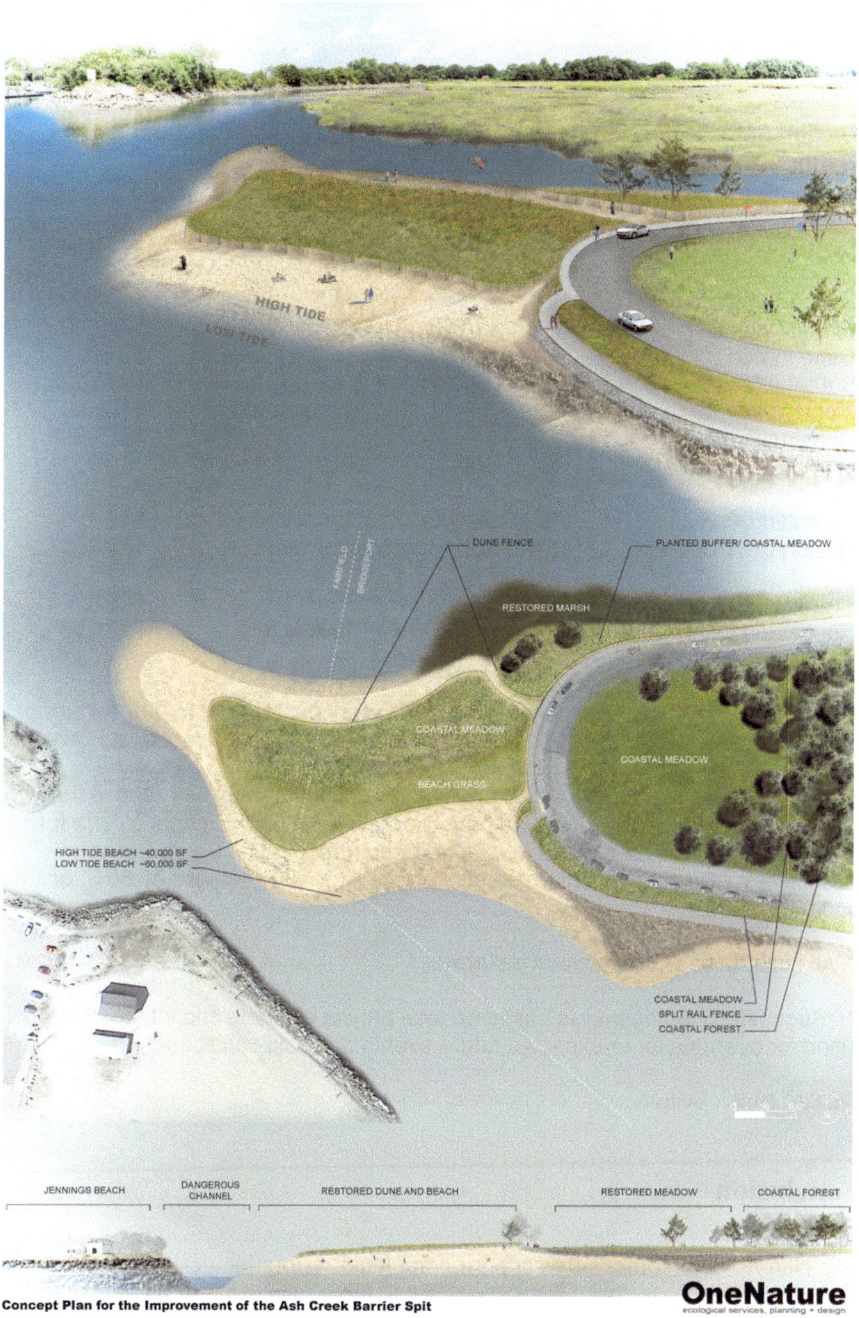

Fig. 8.5 Concept plan for improvement of the St. Mary's-by-the-Sea Barrier Spit

Design Estimate for the Reconstruction and Restoration of Saint-Mary's-By-the-Sea in Bridgeport and Fairfield, Connecticut

Item	Unit Type	Units	Unit Cost	Total Cost
Beach Sand	CY	3,900	$40*	$156,000
Plant Material	Per	42,500	$2	$85,000
Dune Fence	LF	400	$7	$2,800
Subtotal**				$243,800
Mobilization	LS		5% of subtotal	$12,190
Construction Total				$255,990
Permitting	5% of total construction subtotal			$12,190
Design Development	10% of total construction subtotal			$23,380
Total Estimated Cost				**$291,560**
Suggested Contingency***			10% of total	$29,156

All totals approximate. This estimate is preliminary and subject to further revision. This document is for planning purposes only and not intended for construction.

*Cost of sand is assumed for delivery by truck. This cost may be reduced through coordination with nearby dredging projects provided material meets criteria for dune restoration. If "free" sand is obtained from dredging for South Benson Marina or nearby projects then the total project cost will decrease by 50%, or more.

**All construction costs include installation.

***Suggested contingency is based on total project estimate and intended to be used for overruns for unexpected future events and field conditions.

Fig. 8.6 Project Budget

Conclusion

One Nature, with community support, completed restoration design for the barrier spit in 2014. Despite the vast number of benefits that the St. Mary's sand spit provide to the Ash Creek Estuary and surrounding communities, the project to restore and protect the area remains unfunded. If this unique place is not protected soon, the repercussions to the local ecology and community will be long felt. The health of this ecosystem is inextricably linked with current and future benefits that people can enjoy.

References

1. Quinn B, Danzer S (2013) Ash Creek Ecological Master Plan. One Nature LLC
2. Quinn B, Ngoyen C (2014) An Ecological History of the Ash Creek Estuary. One Nature LLC
3. Lathrop, Cornelia Penfield (1930) Black rock seaport of old fairfield Connecticut 1644–1870. The Tuttle, Morehouse & Taylor Company
4. Waldo Jr. GC (1917) History of bridgeport and vicinity, vol 1. S. J. Clarke Publishing
5. Niering WA, Roman CT, Warren RS (1984) Salt marsh vegetation change in response to tidal restriction. Environ Manag 8(2):141–150

Author Biography

Bryan Quinn—Ecological Designer Bryan Quinn is the founder and principal of One Nature, a small multidisciplinary firm located in Brooklyn and the Hudson Valley. Technically trained in restoration ecology, environmental philosophy and landscape architecture, his vision is a world where our development activities improve the environment, not destroy it. He leads a small team of environmental professionals dedicated to building, designing, and growing projects that heal the planet. His work uses ecological knowledge, construction experience, and design principles, in projects that feature designing and constructing research-backed, pragmatic, and compelling places. A recent example involved permitting, designing, and constructing a new half acre civic space in downtown Newburgh, NY using construction methods more typically employed in ecological mitigation projects. They also have a two-acre pesticide/herbicide free native plant nursery grown completely from bioregional genetics..

He holds a Bachelors degree from Knox College and a Masters degree in Landscape Architecture from Rhode Island School of Design, where he was a two-time winner of the prestigious Athena Award of Excellence.

http://www.onenaturellc.com/information/.

Chapter 9
Plan for the Future: Adaptations by Marine Organisms and Shoreline Substrates to Coastal Conditions, Lessons for Planning Human Communities

Marcha Johnson

Nature has provided inspiration for art, architecture, and engineering as long as humans have expressed themselves. In considering how we can learn to live better, more richly and with more care of the coastal systems surrounding cities, natural features have much to teach.

Some marine organisms employ strategies and architectural structures which allow them to thrive in the environments of rapid physical, thermal, chemical, and pressure changes typical of coastal systems. This chapter provides a sampling of the numerous and varied mechanisms which may have aspects that could be applied to human-built infrastructure, buildings, and coastal structures (Fig. 9.1).

Vertical seawalls lead to more violent splashing, waves breaking, turbulence at the toe than porous, gently sloping marshes, and beaches. Consequently, many of the "problems" associated with water undermining walls, overtopping structures intended to keep buildings dry, and erosion are a result of placement (location within wet-prone zones) and shapes that resist natural processes such as longshore

M. Johnson (✉)
ASLA, Brooklyn, NY, USA
e-mail: marcha.johnson@gmail.com

M. Johnson and A. Bayley (eds.), *Coastal Change, Ocean Conservation and Resilient Communities*, DOI 10.1007/978-3-319-41914-5_9

Fig. 9.1 Waves break forcefully against vertical features such as seawalls

drift. New proposals for massive breakwaters with footprints that will kill all benthic organisms and prevent their reestablishment, converting the bay bottom from a flexible feature which supports and feeds beaches, to a rigid underwater barrier, are an example of the kind of well-intentioned but misguided efforts to allow people to think they can remain in low-lying locations along the shore without getting wet during the next big storm.

Some coastal structures are necessary for desirable and important human activities. However, some create unnecessary secondary disturbances which could relatively easily be mitigated. Perhaps the observation of natural adaptations to violently turbulent and harsh seaside environments holds lessons for cities to adapt better to the ocean's edge AND conserve oceanic ecosystems at the same time.

The "Architecture" of Shoreline Organisms [1]

Organisms which thrive in the turbulent, exposed, storm-swept environments of rocky marine waters have adaptations which enable them to endure extreme physical stresses. Intertidal shellfish, for example, can withstand very strong pressures with relatively thin materials. Human-built structures such as massive stone walls, reinforced concrete, and steel sheeting contrast strongly with the flexible strips of eel grass and thin but well-reinforced hard shells of scallops, clams, and mussels. Some live in colonies shaped in ways that protect the soft-bodied creatures within forms that shed the force of waves aerodynamically.

Some of the mechanisms which shellfish use to withstand wave forces include:

- Folded, pleated, or ribbed/corrugated surfaces. Like corrugated cardboard, the folding adds strength to otherwise fragile materials (Fig. 9.2).
- Overlapping plates are reinforced along their edges (Fig. 9.3).
- Double walls result from closely packed colonial shellfish, efficiently conferring extra protection (Figs. 9.4 and 9.5).
- Spiral internal infrastructure produces a very strong core (Fig. 9.6).
- Laminar flow is smoother around round forms and with less resistance, thinner walls suffice (Fig. 9.7).
- Rounded footprints and conical or humped forms. When the widest part is pressed to the substrate, wave forces are distributed to the strongest part of the shell structure (Fig. 9.8).

Sessile, soft-bodied sponges, anemones and tunicates on the other hand, have no shells at all, but instead employ other strategies such as porosity, flexibility, elasticity, and rapid regeneration of bits ripped off by strong currents or being scraped by ice. Plants of the intertidal zone including algae, have another suite of adaptive strategies including:

- Leathery, desiccation-resistant coverings
- Rubbery, flexible and tough yet elastic materials that can adapt to shifts in wave direction and intensity without breaking and minimize heat and desiccation stress

Fig. 9.2 Shell of the angel wing clam

Fig. 9.3 Sketch of overlapping plates

- Knobby, dimpled surfaces such as observed on wrackweed, reinforced by the presence of embedded, thickened nodules
- Frond-like forms attached by holdfasts, can float and move in the waves

Fig. 9.4 Colony of shellfish

Fig. 9.5 Colony of shellfish

Fig. 9.6 Spiral shell

Fig. 9.7 Rounded, streamlined form

- Mat-forming growing patterns such as *Spartina* exhibits, that resists being disengaged from an interwoven root mass, similarly to the way a woven fabric can be pulled in different directions without ripping [2].

Response to Physical Disturbances

Physical disturbances along the coast include crashing waves which move beach sediments, tumbling loose rocks and gravel which grate the shore, and large, coarse, floatable debris such driftwood. These all gouge into the sand along the strand line. Scour by thick plates of sea ice are among the typically rapid, violent disruptions

Fig. 9.8 Shells of sea snails including the (*left to right*): ribbed limpet, shield limpet, masked limpet, and plate limpet

that create the wide variety of microhabitats along shorelines. The mix of recently disturbed unstable/soft soil, hard substrates, novel structures (such as a new pile of driftwood including leafy branches) and intact and stable, vegetated sections, with every intermediate stage of ecological recovery, is one of the reasons for the extremely high productivity of intertidal areas. Organisms adapted to these environments illustrate several ways which seaside towns might be arranged.

The spatial arrangement of organisms on rocky shores exposed to strong, frequent disturbance which continually scrubs away shellfish and seaweed from hard substrates, parallels the spatial arrangement of urban development which is best capable of co-existing with nature. Except where settlement in cracks and other "refuge habitats" [3] gives shelter from exposure to scouring forces, settled organisms (or houses) are periodically physically removed, opening up space for new recruits, including invasion by competing subordinate rapid colonizers (Figs. 9.9 and 9.10).

Urbanized coastal areas share a number of significant physical and procedural parallels with rocky shores: a hard interface with the ocean, a wide range of turbulent disturbances from different and/or unpredictable directions (for example, ferry wakes) and powerful storms such as hurricanes at irregular intervals.

The ecosystem services generated by the oceans are immense, including providing at least 50 % of the earth's oxygen. [4] "Work on [the biology of] rocky shores has elucidated the roles of consumers, competition, disturbance and recruitment in communities" [5] which could ALMOST translate to a description of coastal cities. Probing these comparisons yields a variety of ideas for fitting urban structures and site planning to their coastal setting. Finding ways which would reduce exposure of human populations to storm waves, high water events, and degradation of structures by abrasion and scour WHILE respecting and conserving

Fig. 9.9 Dense populations of mussels in upper intertidal rocky shore can avoid predation by lower-elevation starfish

Fig. 9.10 Many coastal towns traditionally concentrated urban development well above the waterline, avoiding shoreline actions which can displace structures

the primary oceanic character which is required to sustain marine life, as well as human life, will help us evolve improved kinds of cities, ones that abide "by the rules of the sea." [6].

While many natural landscapes are patchy mosaic patterns, an easily distinguishable and immediately noticeable feature of rocky shores is the organization of species in lines parallel to the shore contours, of different colors and textures. This zonation results from competition for light, crowding, response to predator pressure. Shellfish for example, including barnacles, mussels and oysters, sort or distribute themselves along sloped, hard, rocky substrates. There are some advantages and disadvantages to dense crowds of mussels; the presence of barnacles influences the presence of mussels and their predators—starfish. Crowded barnacles on shores with abundant nutrients have the advantages of being supported by neighbors' shells, so they can be more efficient with thinner walls, the way townhouses sharing common walls use energy and space more efficiently than free-standing single family houses.

There may be lessons for coastal cities from the ways shellfish organize themselves in response to stresses. Where resources for human populations are abundant—such as jobs, food, and housing—crowding together at higher elevations on the shore has many advantages. Whereas lower down in elevation, closer to the water, sparse, mobile populations can exist with easily rebuilt infrastructure after floods, limiting exposure to damage from storms and high water. As cities experience the anticipated frequent disturbance and destruction by natural forces, coastal real estate will open up in ways which resemble the openings left after starfish have preyed on densely packed mussels, leaving space for new recruits on the rocky substrate.

There are many habitable inland and upland areas for coastal cities to migrate to over the next half century or so as sea level rises to cover much of the lower coasts. There are also a number of ways for coastal cities to adapt in situ, such as by elevating structures on decks and being flexible and prepared to leave during high water events. Nature illustrates a range of mechanisms, materials, and strategies for living with the ocean's processes. We hope to learn from observing these examples to be better neighbors with the ocean, following the "rules of the sea" to benefit and restore the ocean, as well as allowing us to live within its reach.

References

1. Johnson M (2007) Designing the edge. NYC Parks and Metropolitan Waterfront Alliance, unpublished report
2. Johnson M Designing the edge, pp 24–25
3. Bertness MD (1999) The Ecology of Atlantic Shorelines. Sinauer Associates Inc, Sunderland, MA 227:178–228
4. How much do oceans add to world's oxygen? http://earthsky.org/earth/how-much-do-oceans-add-to-worlds-oxygen. Accessed 26 Mar 2016
5. Bertness MD The Ecology of Atlantic Shorelines, p 228
6. Bush D, Pilkey OH, Neal WJ (1996) Living by the rules of the sea. Durham and London, Duke U. Press

Author Biography

Marcha Johnson, ASLA—Landscape Architect, Ecological Restorationist and Adjunct Professor Marcha Johnson is a landscape architect for New York City Parks, focused on ecologically rich urban waterfronts. She has specialized in landscapes that interface with urban infrastructure with the ecosystems of estuaries, littoral zones, and beaches. An adjunct professor in City College of New York's Spitzer School of Architecture since 1991, she teaches courses in the landscape architecture and sustainability programs. She has lectured at regional and international conferences on her work addressing the incorporation of floodplain functions in urban waterfronts, adaptation to sea-level rise in the context of conserving living ecosystems of coastal cities and recognizing the ecological benefits of novel, spontaneous plant communities. She is currently working on a phytoremediation demonstration project on Randall's Island in the East River, as a more sustainable alternative response to dealing with contaminated "historic urban fill" of urban waterfronts than off-site disposal in distant landfills. Marcha Johnson's written contributions to this book are her own ideas and opinions, independent of policies of either the City of New York or City College of New York.

She holds a B.S. in Biology from U. of Illinois, a Masters of Landscape Architecture from Louisiana State U., and a Ph.D. in City and Regional Planning from U. of Pennsylvania.

Chapter 10
Plan for the Future: Advancing Ecological Engineering Through Science, Conceptual Design, and Community Resilience Building

Adam W. Whelchel

Introduction

By the latter part of the twenty-first century, the globe and all societies present will be faced with compelling environmental and climatic circumstances requiring hard choices and tradeoffs to maintain standards of living and viable environmental conditions. In order to shift that reality, there is a need to think forward today to ensure we have the flexibility to accommodate uncertainty and avoid those costs of inaction. One clear opportunity is presented via a greater understanding of the services and benefits provided to humanity by ecosystems: clean air, clean water, food provisioning, cultural identity and most immediately, a cost effective means of regulating the impacts of natural disasters and climatic change.

According to the United Nations, the economic losses from natural disasters (1970–2013) globally topped \$2.8 trillion with over \$473 billion attributed to earthquakes and tsunamis and \$484 billion to flooding and storms [1]. Globally, coastal zones are particularly vulnerable to flooding due to continued high levels of development and escalating sea-level rise projections [2–4]. In 2005, flood losses for 136 of the world's largest cities was estimated at US\$6 billion per year with projected increases up to US\$52 billion by 2050; with sea-level rise and subsidence factored in, the avoidance costs may reach US\$1 trillion per year [5]. Staggeringly, in 2010 the world's gross domestic product (GDP) exposed to tropical cyclones was 4.47 %, more than US\$1.9 trillion. Since 1970, economic losses from such impacts increased 262 % for high-income countries [6]. Clearly, the difficult lessons of risk and investments have not been heeded in the developed countries. Most disturbing, however, is the growing impact to humanity; for example, an estimated 66 million children around the globe are affected by disasters every year [7–9]. Despite these

A.W. Whelchel (✉)
The Nature Conservancy, 55 Church Street, Floor 3, New Haven, Connecticut, USA
e-mail: awhelchel@tnc.org

M. Johnson and A. Bayley (eds.), *Coastal Change, Ocean Conservation and Resilient Communities*, DOI 10.1007/978-3-319-41914-5_10

alarming trends and forecasts, societies, particularly in coastal areas, continue to aggressively discount disaster risk when making development and redevelopment decisions [10]. The prognosis for people and places in close proximity to waterways is dire without affirmative action to reduce risk by capitalizing on shared strengthens such as ecosystems.

Fortunately, there is stronger global interest and demand than ever before for promoting and advancing resilience to natural disasters and climatic change: as evident by the recent Sendai Framework for Disaster Risk Reduction, new International Council for Science Sustainable Development Goals, United Nations Framework Convention on Climate Change and recent Paris Agreement, and the potential of the World Bank's Adaptation Fund [10–13]. As a global thematic platform, the Partnership for the Environment and Disaster Risk Reduction (www.PEDRR.org) is leading the promotion of ecosystem-based disaster risk reduction implementation and mainstreaming it in development planning at global, national, and local levels [14]. These frameworks, goals, funding, and partnerships all support the use of ecosystems to reduce risk as a recognized and viable path forward. It is clear that as a global society we are on track for an unprecedented risk deficit later this century that will require the employment of every option, particularly ecosystems, given their life sustaining services and co-benefits.

The massive amount of committed and needed investments for infrastructure upgrades globally also presents an immediate opportunity for incorporating the capacity of ecosystems and natural resources (i.e., "natural infrastructure" when used here in the context of risk reduction) into engineered approaches (i.e., referred to here as "Eco-Engineering") to help save lives and make people and communities safer. Estimates by the Organization for Economic Cooperation and Development [15] suggest that US$189 trillion in infrastructure investments will be required globally by 2030. An estimated US$3.6 trillion is required by 2020 to bring infrastructure in the United States up to adequate standards [16]. Investments in infrastructure by governments is typically viewed as a means to promote economic recovery with significant private funding. Public–private partnerships offer a tremendous focal point for dialogue on the capacity of natural infrastructure largely because it can help reduce the level of financing required to meet these projected infrastructure needs in the twenty-first century.

In light of all these circumstances, ecosystems are being increasingly viewed as a critical asset that can help reduce risk to society from disasters and climate change, particularly in coastal areas [14, 17–19]. The protective services along with the co-benefits provided by coastal ecosystems, when recognized, can and are starting to be integrated into comprehensive risk management planning and resilience actions [18, 20, 21]. These recent advancements are supported by a growing body of scientific evidence that coastal habitats such as mangroves, coral reefs, salt marshes, and oyster reefs can reduce the impacts of extreme weather events, storm surge, and flooding [18, 21–25]. Most compelling, however, is that private and public sectors are beginning to prove that coastal habitats are in fact cost-effective defenses [19, 26]. This resonates in watershed management applications where ecosystems such as floodplain forest and wetlands, public amenities (i.e., parks, recreation fields, open space), and agricultural lands are critical assets that reduce downstream flooding risks to urban centers situated along major rivers and receiving estuaries [27].

While the science, management, and economic viability of utilizing ecosystems to help reduce risk is demonstrating promise, there remains a need for further definition, calibrating, designing, testing, and monitoring of natural infrastructure in more situations and at various scales and geographies. In addition, there is a need to create some equivalencies and assurances around the utility of natural infrastructure within the world of engineering and business if we are to truly fuse "ecosystems" with "engineering" and design. To enable a legitimate comparison and eventual adoption of natural infrastructures alongside or integrated with typical engineered approaches, a great deal of multidisciplinary collaboration will be required. The translation of ecosystem dynamics, characteristics, and requirements into guidance, standards, and criteria to allow for regulatory acceptance and permitted use is certainly an immediate goal while the longer term outcome is the mainstreaming of eco-engineering globally during the first half of the twenty-first century. The following discourse provides an overview, conceptual framework supported by case studies, and deeper considerations, to help move this much needed, yet, emerging field of eco-engineering forward.

Ecosystems Defined

To properly conceptualize, implement, and maintain natural infrastructure for disaster risk reduction (DRR) and climate change adaptation (CCA), there is a need to understand and appreciate the complexities of ecosystems through a basic working definition. Ecosystems are essentially integrated and interdependent systems that sustain life. They are in fact living systems which is a critical consideration for eco-engineering. Ecosystems also exist and operate at various interlocking scales from the local embayment to an entire estuary and coastline (Fig. 10.1); from a floodplain forest to an entire watershed; from a single hillside to an entire mountain range. Most remarkably, ecosystems shift and adapt to a diverse array of external and internal forces over time.

Ecosystem Characteristics

There are certain characteristics of ecosystems that absolutely need to be considered prior to contemplating their use to reduce risk and improve resilience for society. Ecosystems, large and small, are driven by dynamic and ever changing climatic and

Fig. 10.1 Intact natural coastline on Campanario Point and Cano Island Biological Preserve in Costa Rica's Osa Peninsula. © The Nature Conservancy/Sergio Pucci

physical forces that may include waves, wind, rain, storms, drought, heat, slope, and geology. In most ecosystems it is the unique combination of these forces (at various magnitudes) that define current and future form and function. For example, the width of naturally occurring salt marsh is often determined by waves and wind along the coast as well as precipitation events that carry upstream sediment to marshes in deltaic systems (i.e., Mississippi or Rhine River Deltas). Given this reality it becomes imperative to recognize that there is uncertainty in the future form of ecosystems. Ecosystems can and are adapting to ever changing conditions; be it from natural forces and/or disturbances by mankind. In addition, ecosystems require a great deal of space and often cannot be forced into places for unintended purposes predefined by a project scope; regardless of design prowess and engineering capabilities. Finally, ecosystems often react in unexpected and non-confirming ways in response to site conditions and forces. In contrast, typical default engineering approaches in coastal (Fig. 10.2), riverine, and mountainous locations result in static structures hardened against specific forces (waves, wind, precipitation) that are singular in purpose with specific design life and predictable reactions to stress. A balance between form and function with careful considerations of limitations is needed to truly integrate ecosystems and engineering and design (i.e., eco-engineering).

Fig. 10.2 Static coastal engineered structure hardened against localized waves, wind, and currents. © The Nature Conservancy

Importance of Ecosystems to Mankind

Unlike typical engineered structures in coastal zones, ecosystems provide a diverse array of ecosystem services and co-benefits [28, 29]. Provisioning services that are advantageous for mankind include clean air, clean water, sustenance, and livelihoods for much of the world's population. For example, an estimated 56.6 million people work in fisheries and aquaculture (2014); while in 2013 more than 3.1 billion people relied on fish for 20 % of their average per capita animal protein intake [30]. Regulating services include buffering communities from routine and extreme weather and changing climate that affect health and wellbeing (Fig. 10.3). For example, along the eastern coast of the United States, coastal wetlands were estimated to provide approximately US$23.2 billion per year in storm-regulating services [31]. Cultural services include social identity, spiritual replenishment, public amenity, and recreation. In addition, ecosystems provide habitat that sustains life. In total, ecosystems provide a much more attractive and value-added proposition to securing resilience versus typical engineering approaches such as seawalls and dikes. With greater recognition of these ecosystem services and co-benefits, risk reduction and resilience stand a greater chance of achieving a net positive for society [32].

The single greatest natural infrastructure-based strategy for reducing risk to mankind is to ensure that exiting ecosystems remain intact and undisturbed in perpetuity. To ensure these systems remain viable, proactive steps should be taken

Fig. 10.3 Conserved dunes and salt marsh dissipating wave energy (*lower right corner*) versus seawall around residential property (*upper right corner*) along eastern seaboard of United States during Tropical Storm Sandy. © Adam Whelchel

so that future development or redevelopment do not impact ecosystems and are out of harm's way from hazards such as flooding and landslides. At the very least, the societal benefits from and cost of losing ecosystem services should be carefully considered in planning processes.

A growing number of affirmative examples in terrestrial, freshwater, coastal, and marine ecosystems in which services and co-benefits are assessed are helping to inform the design of spatially explicit management plans [33–37]. In the coastal zone of Belize, the explicit integration of ecosystem services into future management scenarios within a national coastal planning process resulted in a preferred option that increases by 25 % coastal protection and doubles fishing revenue [20]. The key as pointed out by these authors, was the collaboration between management and science (including ecologists and engineers) that allowed for the consideration of multiple services and benefits within an iterative, yet robust stakeholder engagement process. An earlier analysis of economic benefits (market-based and value transfer methods) generated from coastal ecosystems in the Bohol Marine Triangle in Philippines revealed compelling enough results [US$11.54 million over 10-years (10 % discount rate)] to be incorporated into spatially explicit management plans to sustain natural assets [38]. These examples encourage further exploration of natural assets and solutions to risk management in situations where the general public, private sector, and policymakers are properly engaged and educated on ecosystem services and co-benefits as viable alternatives or compliments to typical engineered approaches.

Ecosystems in a DRR and CCA Context

What does the science tell us?

The specific question of how effective natural infrastructure are at reducing risk is a rather new exploration for academic, business, and public sectors. However, peoples around the world have been responding to and recovering from hazards using natural infrastructure for millennia. Traditional or indigenous types of risk management have been existing in a "pre-modern engineering" era; relying on flexible approaches to settlement, agriculture, and trade; followed by resettlement. Much of this traditional knowledge about resilience can still be seen in populations that live outside the reach of modern engineering; in rural, sparsely populated areas or in developing nations that lack funding to protect the vast numbers of exposed peoples across large geographies (i.e., Assam, India). Much of the traditional risk responses rely on taking advantage of seasonal weather patterns, respecting the magnitude of extreme weather events, being able to quickly communicate pending risk, relocating and recovering in ways that minimize future impacts, and most importantly, continuously reinforcing awareness through oral history, proverbs, memorials, and the like. A time-honored example of traditional understanding and response to flooding is seen in Japan.

The people of Japan have a long history of dealing with disasters such as earthquakes, tsunamis, and typhoon-induced floods, with their oldest known history books capturing the impacts and responses as far back as 416 A.D. [39]. Positioned in the pacific "Ring of Fire" and appropriately named "Typhoon Alley", the Japanese people have accumulated knowledge and approaches for dealing with extreme natural events reinforced and nurtured by a culture of continuous awareness. Impacts from flooding have been particularly devastating to this island nation for thousands of years due to intense, typhoon-induced rainfall events (mean annual rainfall 1700 mm) and short, steep, and small river basins [40, 41]. Nishikawa [39] traces the origins of modern Japanese disaster risk management to priests sent to China to retrieve Buddhist teaching (sixth to tenth century) and returning enlightened regarding rudimentary civil engineering practices which were quickly applied to mitigating flood risk. One of the first integrated risk management projects in Japan was a dam and reservoir complex built in 731 A.D. in Hyogo prefecture that had the dual purpose of containing flood waters and irrigation for farming; the structure is still in use today. Other research carefully documents shifts in flood management over the centuries with control structures (drains and embankments) dating back to 300 B.C. with systematic design and use beginning to arrive in 1467 A.D. eventually manifesting in the infamous Shingen bank built by the sixteenth century warlord Takeda Shingen [42]. By the seventeenth century, practices shifted focus from embankments and reservoirs to large-scale river channel straightening with adjoining floodplain storage. Interestingly, the use of floodplain storage was replaced in the 1800s by more aggressive channel straightening and higher embankments to ensure the quickest path to the sea. During the 1940–1950s, Japan was influenced by flood control practices in the United States involving construction of large dams. Due to several catastrophic dam failures, a shift toward using floodplains once again became common practice in the 1990s via what is referred to as "Green Dams." [42]. This history illustrates how reservoirs and natural infrastructure such as floodplains and river channels were used, then discounted over time and finally utilized once again in modern times as valuable management assets (Fig. 10.4). Regardless of these shifts one aspect has remained constant in Japan, the tradition and cultural awareness of disasters with reminders of the suffering endured via erected shrines, works of art, and proverbs.

In the context of "post-modern" or current engineering approaches, the need to secure a higher level of confidence and certainty about natural infrastructure as a risk reducer is being taken up in earnest. In the past 20 years, a growing number of efforts around the world have focused on shifting ecosystem restoration and conservation into the context of reducing flooding and erosion control [43]. One of the first synthesis reviews included 69 projects across a variety of coastal habitats including coral reefs, mangroves, salt marshes, and seagrass and kelp beds [44]. The results from this first meta-analysis provide significant assurances on a central concern for people, businesses, and governments in coastal areas globally; the heights of waves as they make landfall. Based on the studies currently available, the authors determined that average reductions in wave height range from 35 to 71 % depending on habitat type and location along the coast. At the top of the list for wave reduction services was coral reefs (70 %) and salt marshes (72 %); lower reductions were observed in mangroves (31 %), and seagrass and kelp beds (36 %). Coral reefs were clearly a standout for wave reduction because they typically occur

Fig. 10.4 Modern flood control and water management reservoir in downtown Tokyo, Japan. © Adam Whelchel

in high energy coastal environments versus salt marsh and mangroves that occupy lower energy locations. While this study suggests some promising reductions in risk with natural infrastructure, the extent of comparable data currently curtails the ability to translate physical characteristics of these habitats into design parameters for engineering purposes with the exception of coral reefs.

In the Caribbean, the country of Grenada is recognized as highly vulnerable to the impacts of climate change with cost estimates expected to reach 111 % of current GDP by 2100 [45]. In hopes of countering this prognosis, a significant eco-engineering project is underway to construct artificial coral reefs to help reduce coastal erosion in a critically important portion of Grenada's coastline—Grenville Bay [3–unpublished]. The core objectives of the Grenville Bay effort are to restore the wave-breaking services of degraded reefs as well as reestablish reefs that have been lost. To do this, the project is testing an engineered submerged breakwater design that has higher ecological values and is less expensive than typical onshore engineered structures (i.e., seawalls, groins). At the project onset, 60 years of wave dynamics for Grenville Bay was modeled along with storm events and detailed bathymetry to aid in the identification of optimal placement, configuration, and characteristics of submerged breakwaters for reducing wave energy and currents. Design requirements for the artificial reefs include:

- Must reduce wave energy passing over the structure by 80–90 %,
- Last for 30-years,
- Be cheaper to construct and maintain than the traditional breakwaters,
- Use local materials,
- Be built by local communities, and
- Support the growth and accretion of living coral.

The result was a new design of welded gabion baskets made with 5/8″ rebar that can be filled, stacked, and locked together in a pyramid formation of various widths and heights depending on specific modeling outputs and site constraints (Figs. 10.5 and 10.6). In January 2015, four pilot reef structures covering 30 m were installed with eventual full buildout to reach 20 individual structures over 300 m. Monitoring of the pilot reefs will provide data to re-inform the pilot design and full-buildout plans. When completed this project will represent the first example of how local communities and governments of small island nations have utilized natural infrastructure coupled with sophisticated modeling, conceptual design, and engineering principals to reduce the impacts of coastal forces and climate change.

Protective services provided by other habitat types such as mangroves can also be pronounced in a variety of categories including provisioning of food and coastal defense from routine waves, storm surge, and tsunamis [46]. Under non-storm conditions mangrove forests are highly effective at reducing waves by approximately 70 % of nearshore height and eliminate scour of the mangrove bed [47]. Other studies have shown that mangrove forest width influences wave reduction; 13–66 % over 100 m versus 50–100 % over 500 m [46]. This is due in large part to frictional drag within the water column by mangrove trunks and roots (Fig. 10.7).

The density, spacing, age, and size of tree, as well as the species of mangrove can also have a dramatic effect on wave height reduction. Under storm surge conditions, there is a reduction of 4–48 cm per kilometer wide forest depending on hydrodynamics in front of and within the mangroves which can have an impressive reduction in landward flooding extent to coastal property [46]. These authors also suggest that local topography of the mangrove forest and in the run up is a driving factor in determining wave reduction during storm surge. Under tsunami conditions, mangroves can reduce flood depths by 5–30 % over 200-meter or more wide forests, however wave heights above 4 m will decimate mangrove forests regardless of width [46]. Under many of these scenarios, mangrove forests are able to decrease the flood level on the landward edge by at least 35 % in the presence of both barrier and fringing reefs in coastal Belize [47].

In more temperate regions of the world, coastal salt marsh is a focal habitat when considering protective services and risk management. Under experimental conditions in a 300-meter wave flume with transplanted natural marsh vegetation, wave attenuation was estimated by Moller et al. [24] at 60 %. These authors also noted that despite the eventual damage to the vegetation the marsh substrate remained resistant to erosion suggesting that salt marsh is an important element for coastal risk reduction. In the Yangtze Estuary, China waves ranging between <0.1 and 1.5 m were effectively attenuated completely over a marsh width of 80 m with the exception of the largest waves which required >100 m [48]. For a recently restored marsh in front of a dike in the Netherlands, van Loon-Steensma et al. [49] demonstrated that under storm conditions (frequency of 5–10 times/year) a 90-meter wide salt marsh will attenuate waves by 80 % while under storm conditions with a frequency of 1/2000 years, waves were attenuated by 50 %. These

Fig. 10.5 Installing artificial submerged breakwater reef structure to help reduce wave heights that cause coastal erosion in Grenville Bay, Grenada. © The Nature Conservancy

Fig. 10.6 Artificial reef structure using welded gabion baskets made with 5/8″ rebar that can be filled, stacked, and locked together in a pyramid formation of various height and width (Grenville Bay, Grenada). © The Nature Conservancy

Fig. 10.7 Mangrove in the shallow coastal salt flats of Warderick Wells Cay in the Bahamas Exuma Cays Land & Sea Park. © Godfrey/The Nature Conservancy

authors suggest that salt marsh when restored in front of a dike will reduce wave attack on the adjoining dike thus enhancing the overall risk reduction of an integrated flood control system. In addition to salt marsh width, Yang et al. [48] also demonstrated an inverse correlation between plant height (*Spartina alterniflora*—tall and short form) and wave attenuation rates; the taller the vegetation the more wave attenuation. Based on these few studies the width, landscape position, and vegetation type are certainly important design considerations for natural infrastructure projects involving salt marsh.

One of the main concerns for researchers, risk managers, and dependent communities is whether or not coastal habitats such as mangrove forests will in fact be able to remain intact and functional as sea levels begin to escalate [46]. In certain locations, sea-level rise projections, landward topography, and the presence/absence of the built environment are being considered using technology to identify potential advancement zones or future habitat locations for salt marsh [50]. In Connecticut (USA), the entirety of the state's coastline (24 coastal municipalities) has been analyzed down to the parcel scale (i.e., finest scale of land-ownership) for a Salt Marsh Advancement Zone Assessment which identifies the future location of salt marsh with down-scaled sea-level rise projections [51, 52]. This study allows for the visualization of potential future conflicts and opportunities for the built environment and natural resource advancement zones, respectively, across a critical urban corridor along the eastern seaboard of the United States (Stamford–Bridgeport–New Haven). Assessments of this type are instrumental in shaping a community resilience building dialogue that incorporates the risk reduction services of existing as well as future habitat in the context of coastal change due to sea-level rise. Whether habitats can advance and maintain remains an open line of research, however, the provision of space (i.e., advancement zones) for habitat to persist within desirable coastal areas is absolutely a prerequisite.

In many tropical coastlines the benefits of habitat complexity where diverse types co-occur results in a reinforcing system of defense (Fig. 10.8). One recent study in coastal Belize examined the individual and combined protective services provided in progression by coral reefs, seagrass meadows, and mangrove forests under storm conditions and sea-level rise scenarios [47]. The study found that while mangroves are clearly the most effective of the three at reducing wave height and energy, living coral reefs and co-occurring seagrass meadows further reduce the risk. Effectively, the presence of all three habitats offers greater protection then single habitats or combinations of any two habitat types. In addition, it has been suggested that the presence of coral reefs and seagrass may in fact increase the viability of adjoining mangrove forests by increasing recruitment [53]. Interestingly, this function of coral reefs and seagrass meadows has been replicated to some extent artificially in proactive mangroves restoration projects in Indonesia [54]. Finally, Guannel et al. [47] strongly suggest that when considering ecosystems in a risk reduction context it is important to take a "whole systems", integrated (multiple habitats and forces simultaneously) approach versus a singular assessment with one habitat at a time. Again this consideration recognizes diverse services provided by natural infrastructure versus typical engineered approaches that are

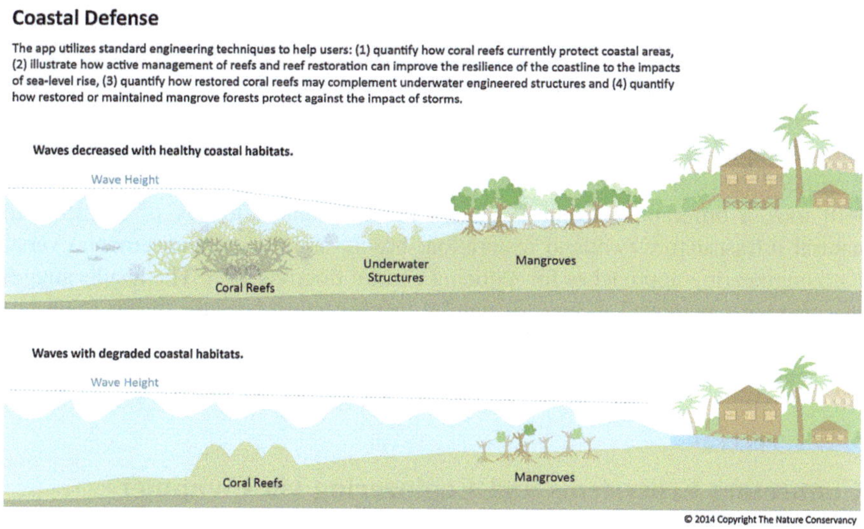

Fig. 10.8 Informational graphics depicting the protective services and co-benefits provided by coastal ecosystems with and without certain co-occurring habitat types such as living coral reefs and mangroves in proximity to people and property. © The Nature Conservancy

often singular in function, form and utility, and lack the capacity to self-repair or regrow stressed sections.

One particularly useful construct with applications for many developed segments of coast around the globe is that of multiple lines of defense [55] where a suite of co-occurring habitat and natural infrastructure are managed in compliment with or at least in proximity to hardened engineered features (i.e., groin field and breakwaters with salt marsh and beaches). This concept of co-managing the use of natural and hard engineering features (nonstructural and structural flood control, respectively) to reduce storm surge as well as advancing habitat restoration goals has been integrated and adopted as a core strategy by the state of Louisiana and associated federal coastal plans in the aftermath of Hurricane Katrina and Rita [56]. Ultimately, co-occurring natural and engineered solutions should be viewed as assets that work in accord toward risk reduction for adjoining communities. When combined with active habitat restoration and consideration of future advancement, improved resilience becomes a discussion around replication and redundancy in form and function of both natural and engineered solutions for a given location. Again, this helps to expand the definition of multiple lines of defense in coastal settings to include redundancy which provides greater flexibility as the intensity, duration, and frequency of storm events and sea-level rise continue to escalate.

Currently, there is unfortunately very little reporting on costs and benefits of ecosystem-based DRR and CCA; in particular comparative studies of natural solutions versus typical hard engineering approaches or hybrid solutions. A few studies (as reported by Shreve and Kelman [57]) in the Philippines estimated that the cost–benefit ratio for a reforestation project for DRR was approximately 30 for a

15-year, 1000-hectare project [57, 58]. The cost–benefit ratio for a typical engineering approach in the Philippines was valued at 4.9 and 0.67 for seawalls and dykes, respectively [59]. Although not directly comparable, the suggested higher risk reduction returns on investments using natural infrastructure in the Philippines is only amplified when sustainability and resilience improvements are linked with livelihood co-benefits (i.e., regulating plus provisioning services). A recent study [60] took a different tact and examined the willingness to pay of residents for natural infrastructure (wetland restoration, beach, and dune enhancements) versus hard engineering approaches for reducing coastal flooding [61]. The results suggest that residents preferred to pay for natural infrastructure because of the flood protective services coupled with the co-benefits (public amenity, aesthetics, recreation) that seawalls simply do not provide.

Connecting Ecosystems and Engineering via Computer Programs

Another central challenge to fully integrating natural infrastructure into risk management practices around the globe is the limited awareness of the tools currently available to aid decision-makers with linking the "eco" with "engineering" [62]. While the number of tools is growing due to the frameworks, goals, funding, and partnerships discussed earlier there are fewer examples of how they have actually advanced decisions involving natural infrastructure and risk management. Without real-world application the utility and value of tools will not be realized and necessary improvement to existing and development of the next generation applications will have little support. One of the obstacles, beyond awareness, is that the variables in play that ultimately determine a successful integration of natural infrastructure are complex and expensive to model with sufficient confidence to direct larger investments. Often times the available tools are at best coarse level planning application that help surface potentially suitable areas for projects that involve risk management. Generally speaking, the tools of today help to present environmental and socioeconomic data geospatially to guide planning with a few that allow for future visioning of sequential action scenarios. In some coastlines where adequate data is available, tools provide sophisticated modeling and analysis of disasters and climate change impacts to both natural, man-made, and hybrid systems (e.g., Delft3D and Mike21). The tools of tomorrow will allow for a spatially explicit, quantitative expression of independent and sequential risk management actions that truly integrate natural infrastructure. Regardless of types and availability of tools, there is no substitute for fusing said tools and information into a collaborative, community-based process that comprehensively considers risk and connect with ongoing resilience management efforts.

As reviewed by Kroeker et al. [63], a principal utility for tools that link ecosystems with engineering in a coastal context is to estimate the protective services by calculating the effects of ecosystems on flooding and erosion—essentially estimates of exposure and vulnerability. These "index-based approaches" allow the practitioner to examine the individual changes in risk through various natural resource management actions. The authors differentiate "process-resolving approaches" as more sophisticated approaches to define and integrate various forcing variables (waves, storm surge, tides, etc.) and processes (sediment transport) to more specifically determine risk and appropriate solutions. This class of tool, however, has requirements (data, expertise, advanced technology, funding) that preclude wide ranging use. While there are a number of important tools for estimating risk reduction from ecosystems (see Chaps. 4 and 5 in World Bank 21), the Coastal Vulnerability and Coastal Protection modules in InVest as expressed in the Coastal Resilience platform are most widely used (18 states in US; 7 nations globally) [21, 63–65].

In coastal portions of the state of Alabama, the Coastal Resilience platform is being used to help advance a ground breaking natural infrastructure project; "100–1000 Restore Coastal Alabama". Centered on the fourth largest estuary in the United States—Mobile Bay (1070 km^2)—this long-term project seeks to restore and enhance oyster reef, seagrass beds, and salt marsh lost due to coast-altering activities including dredge and fill operations and seawall and jetty construction. Anecdotal accounts suggest that much of the historic reefs in the Mobile Bay area were harvested in the early 1900s for road and infrastructure construction. With the reefs removed, the softer sediments supporting seagrass and salt marsh were largely lost due to the unmitigated impact of more intense and frequent wave energy. The project's objective is to build 161 kilometers of oyster reef breakwater (Fig. 10.9) that will help to promote the reestablishment of over 405 hectares of coastal marsh and seagrass (i.e., 100 miles–1000 acres). Once established the intended outcomes include a more resilient and viable coastal ecosystem with reduced erosion and risk to property, improved habitat for commercial and recreational fisheries and shellfish, and an enhanced tourism experience. A variety of engineering techniques and practices have been employed to construct the oyster reefs using concrete, rebar, and oyster shell in configurations referred to as bagged shell, reef balls, and oyster castles (Figs. 10.10 and 10.11). The anticipated 10-year, US$150 million investment is expected to generate over 300 jobs per year, elevate regional household income by US$9.7 million, increase annual commercial fisheries revenue by US $6.87 million, and most importantly, provide an effective natural infrastructure alternative to typical engineered approaches such as bulkheads for residents of Mobile Bay. This project represents an affirmative use of science and natural infrastructure to help restore the estuary and improve the resilience for municipalities, businesses, and residents dependent on this coastal resource.

With over 150 kilometers of coast in Mobile Bay and a need to effectively reduce risk to that perimeter, the "100–1000 Restore Coastal Alabama" project is utilizing the Coastal Resilience (www.coastalresilience.org) decision-support platform. This tool provides a web-based suite of map layers and apps via an intuitive, user-friendly platform that allows for the integration of natural infrastructure into disaster and climate change planning [62, 66, 67]. The tool spatially defines risk

Fig. 10.9 Installed "bagged-shell" oyster reefs shown reducing wave height and dissipating energy along exposed coast in the Gulf of Mexico, United States. © Constantini/The Nature Conservancy

reduction characteristics of natural infrastructure such as oyster, mangroves, and coral reefs within DRR and CCA contexts in tandem with critical socioeconomic considerations at local to national scales. Within the Coastal Resilience platform resides the Coastal Defense Application which is being used in coastal Alabama to identify the protective services from existing reefs and wetlands as well as allowing the user to interactively locate and design new natural infrastructure projects in the context of both local and larger scale risk management strategies. The application uses the model InVEST to incorporate standard engineering approaches to calculate reductions in wave height and energy due to natural infrastructure with user-defined wave and surge characteristics, sea-level rise projections, and the location of existing coastal habitat and hard engineering structures [65]. The "100–1000 Restore Coastal Alabama" project provides an important fusion of technology, natural resource management, and engineering to proactively and effectively reduce risk from storms, flooding and sea-level rise in a dynamic coastal setting.

The science and tools currently available suggest there is added value in considering and integrating natural infrastructure into the overall approach to coastal risk management going forward. The current state of knowledge compels coastal communities and governments to incorporate the diverse services and co-benefit provided by ecosystems including risk reduction. This information provides the "what" and "where" in terms of potential solutions and begins to highlight the physical and cost effectiveness. What is needed, however, is further clarity of how

Fig. 10.10 Community-based installation of concrete oyster castles (hybrid solutions) to reduce wave height, minimize shoreline erosion, and increase ecosystem function of coastal resource in Mobile Bay, Alabama. © Nortemann/The Nature Conservancy

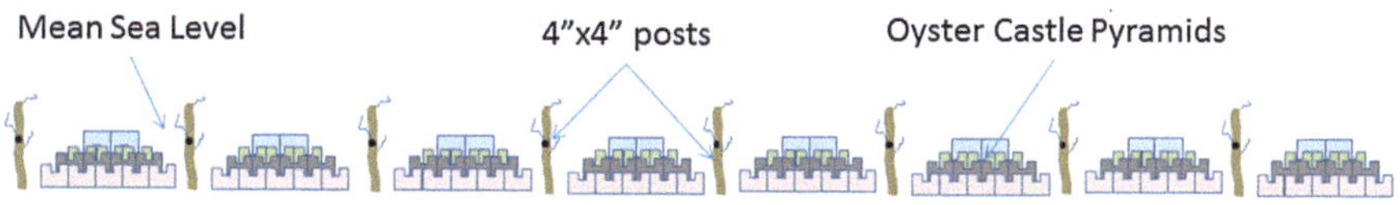

Fig. 10.11 Cross-sectional profile of oyster castle pyramid series used to reduce wave height and minimize shoreline erosion and support the establishment of salt marsh and seagrass habitat landward of these in-water, intertidal structures. © The Nature Conservancy

to integrate this into useful and useable design ("eco-design") and engineered approaches that resonate with individuals, communities, corporations, and governments trying to reduce risk and improve resilience (Box 10.1).

Box 10.1 Key Recommendations to Enhance Natural Infrastructure Acceptance and Applications Through Science, Tools, and Engagement (Modified in Part from World Bank [21])

- Focus natural hazard mitigation and master plans on improving the resilience of existing habitats and ecosystems as a cost-effective approach to risk reduction at all scales.

- Seek improvements to ecosystem service models and tools that allow for a more robust accounting of protective services and co-benefits in risk management planning including existing natural resources. Compare projects outcomes with and without natural infrastructure. Include estimates at the project specific and national scale.
- Consider restoration, enhancement, and/or creation of habitats as key alternatives in risk management plans and actions.
- Advance exploration of hybrid design and applications in diverse situations with long-term monitoring programs.
- Implement additional pilot projects across a variety of coastal conditions and broad geographic extent to fully demonstrate the effectiveness of natural infrastructure in reducing risk as compared to hard engineered approaches. Ensure that rigorous experimental design and long-term monitoring is integral to new pilot projects.
- Work to develop of natural infrastructure standards and criteria acceptable to the engineering and design communities as well as the regulatory/permitting agencies.
- Conduct robust community-based engagement around the benefits of natural infrastructure. Particularly include coastal and riverine property owners.
- Convene and participate in interactive dialogue between engineering, design, policy, and natural resource management professionals that continue to explore the efficacy of natural infrastructure, hybrid approaches, and typical engineering practices.

Integrating "Eco" into "Engineering"

Calibrating the Terms

Before moving forward with "how" to incorporate ecosystems as an important element in risk management it is necessary to categorically define what is meant by eco-engineering. Clearly, the world is full of examples where engineering is singular in purpose and does not consider the ecological needs of the natural systems being altered or manipulated (i.e., "non" eco-engineering). The most pronounced of these include large-scale manipulation of river systems through the building of dams and attempts to stabilize shorelines through the installation of massive seawalls (i.e., Shanghai, China) (see Fig. 10.2). In many coastal areas such as in Europe, typical engineered structures (seawalls, groins, breakwaters, etc.) have increased the loss of marsh and their protective services, thus increasing exposure of society to storm impacts [68]. These types of coastal structures are arguably limited in utility and rarely cost effective when full-cost accounting that includes ecosystem services and co-benefits are factored in. Multidisciplinary, comprehensive approaches that have a much broader scope will ultimately lead to greater creativity when addressing past engineering mistakes and preparing for a climate-changed future. With this frame, incorporating the services and co-benefits provided by ecosystems in multi-objective planning for risk reduction projects is imperative.

At the same time these "non" eco-engineering examples are considered, it is equally constructive to examine "false" or "pseudo" eco-engineering. These are engineered approaches that seek to incorporate elements of ecosystems and do so in ways that resemble aspects of ecosystem form, but rarely the ecological functions. These attempts usually capture some superficial appearances and/or the aesthetics of ecosystems but fail to fully recognize the flexible, adaptive characteristics of ecosystems particularly in the context of dynamic responses to disturbances and changing conditions. A prime example of "pseudo" eco-engineering are the artificial islands on the coast of Dubai, United Arab Emirates. The recently completed Palm Jumeirah (2014) and the projected Palm Jebel Ali will eventually represent approximately 530 square kilometers of private residences placed on sandy fill dredged from the adjoining Persian Gulf and protected by approximately 6.4 million kilograms of stone revetment (no concrete or steel was used). This development attempts to mimic the aesthetic beauty of coastal beaches but with direct impacts to existing habitat (94 million cubic meters of fill) with sophisticated engineering and placement of hard engineered features in an attempt to stabilize a previously dynamic coastal ecosystem. Inevitably, ecosystem processes such as coastal erosion, wave patterns, and sediment transport are altered by this type of development.

Categories and Typologies for Eco-Engineering

To fine tune the universe of eco-engineering options (excluding "non" and "pseudo"), a constructive enterprise is to first categorize viable types. Broadly speaking, there are three useful categories of eco-engineering: (1) Existing/Intact ecosystems, (2) , and (3) Hybrid solutions. Existing and intact ecosystems provide inherent protective services and co-benefits yet are not specifically managed for disaster risk reduction and climate change adaptation. This type occurs naturally and has escaped the direct negative impacts of development or mismanagement to date. Natural infrastructure systems are created or restored and managed with the specific purpose of risk reduction along with associated co-benefits and objectives. The final type—hybrid solutions—are comprised of natural infrastructure (non-structural) with engineered structures (Fig. 10.12). This could be the cumulative suite of natural and engineered applications in one general location or it could encompass engineered features with natural features integrated within a project (i.e., living shorelines) (Figs. 10.13 and Fig. 10.14).

With these three broad categories in mind, there is an additional consideration: a need to develop functionally comparable equivalents between natural infrastructure and typical engineering approaches or typologies. By reviewing the full spectrum of approaches in a simple side-by-side fashion for a given location or project, including a comparison of negative impacts and limitations, can facilitate dialogue about different forms, yet similar functions. For example; a visualization of an existing/intact ecosystem (see Fig. 10.1), natural infrastructure, and a hybrid solution (Fig. 10.12) paired against a typical engineering approach (see Fig. 10.2) for different coastal zones and conditions (i.e., onshore, near shore, off shore, shallow, deep, dynamic, static, etc.) provides a useful forum through which coastal engineers, ecologists, and planners can develop comparability across categories and typologies for risk management using natural assets (Fig. 10.15). For example: offshore living breakwater (Grenville Bay) versus a hard engineered breakwater; nearshore oyster castle with adjoining salt marsh (coastal Alabama) versus an onshore seawall or revetment.

Eco-engineering categorization and development of typologies serve as a precursor to introducing conceptual design as a progressive step toward more sophisticated project and site planning that builds in existing/intact ecosystems, natural infrastructure, and hybrid solutions. Conceptual design, if used appropriately and fused with a community resilience building process, can be a powerful

Fig. 10.12 Hybrid solution ("living shoreline") in low energy embayment (Niantic River, East Lyme, Connecticut) with retaining wall, low profile vegetated dune, and sandy edge. Dock is not tied to embankment which minimizes scour to shoreline from tides and ice and provides space for natural infrastructure. © Adam Whelchel

Fig. 10.13 Transformation of shoreline from engineered to hybrid solution by integrating existing wooden/steel bulkhead and installing earthen filled, fabric compartments to allow for stabilization and establishment of vegetated embankment along tidal creek in Mobile Bay, Alabama. Pictures follow in chronological order from left to right: May 2014, early June 2014, late June 2014, August 2014, March 2015 © The Nature Conservancy

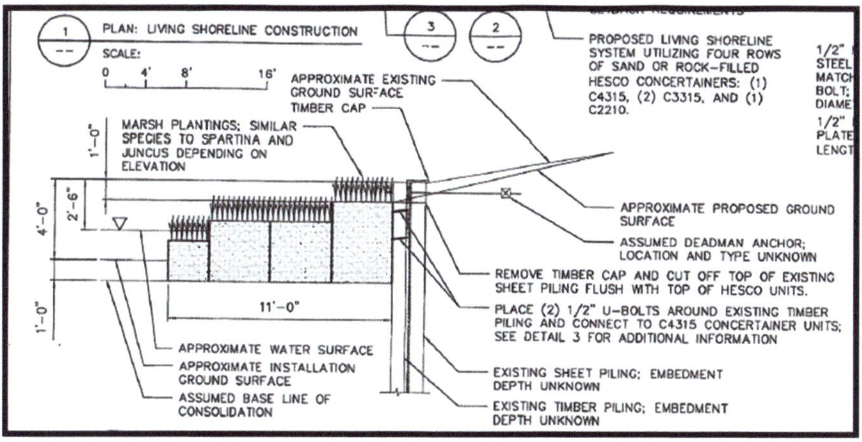

Fig. 10.14 Engineering plans and specificationsfor living shoreline (hybrid solution) construction in Mobile Bay, Alabama (see Fig. 10.13 for as-built project). © HESCO Bastion Environmental Inc./TNC

motivator. A robust conceptual design process allows for collaborative input from the community to help shape the future vision for a variety of locations, from activating an industrial waterfront parcel, to planning a large port facility, to considering the entire coast line of a region. The process can also help to educate stakeholders on the net positives of using natural assets to reduce risk ("eco-design"). Properly facilitated, conceptual design allows stakeholders to visualize the future and creates buy-in with positive outcomes.

Fig. 10.15 Comparison of coastal typologies using typical engineering (*upper left*; *lower left*) and eco-engineering practices and techniques (*upper right*; *lower right*). © The Nature Conservancy

Pathways Forward for Eco-Engineering: Community Engagement and Conceptual Design

One of the most important steps in advancing natural infrastructure as a viable option for risk reduction is community engagement. In particular, Community Resilience Building (www.communityresiliencebuilding.com) has proven to be highly effective in bringing eco-engineering and design applications into a comprehensive and collaborative, "whole" community approach to improving resilience. At the core of any community engagement effort like the Community Resilience Building (CRB) Workshop process is the support of a diverse array of stakeholders before, during, and after planning to activate champions and secure community momentum. It is usually necessary to go beyond the typical emergency management professionals to others including major employers, health care providers, regional to local planners, neighborhood representatives, coastal engineers and designer, elected officials, and natural resource professionals to name a few. In CRB Workshops this broad group of participants develop hazard, ecosystem, infrastructure, and societal profiles of their community using a core tool known as the Risk Matrix along with participatory-mapping exercises (Fig. 10.16). The Risk Matrix enables the community to collaboratively catalogue vulnerabilities and strengths such as wetlands, beaches, and floodplains in an interactive dialogue that allows for interdependent risk management situations to be verbalized (i.e., chronic storm runoff on a road surrounded by protective wetlands that services a daycare facility for children). Workshop participants then list all actions to reduce vulnerabilities and reinforce strengths for each community profile followed by a relative prioritization of actions over time against all top hazards. The CRB Workshop process can accommodate all hazards (e.g., storm surge, wildfire, tornados, drought, sea-level rise, landslides, tsunamis, earthquakes, etc.) in any circumstance (e.g., coastal, urban, rural, watersheds, deserts, high elevation, developed and undeveloped) across all governance and societal structures and at any geographic scale (Fig. 10.17). Experiences with community-based engagement in over 35 communities to date, reinforce that managing coastlines is about protecting people, property, and places. Without robust community engagement there is always a concern that critical knowledge will not be incorporated into plans and actions taken may encounter resistance to efforts perceived to cause more harm than good.

As a precursor to a larger regional effort, the CRB Workshop was recently used with coastal municipalities in Connecticut (USA) to increase receptivity and secure

Fig. 10.16 Community Resilience Building Workshop participants collaboratively share experiences and comprehensively create priority actions using participatory mapping and the Risk Matrix. © Adam Whelchel

Community Resilience Building Workshop Risk Matrix (www.communityresiliencebuilding.com)

H-M-L priority for action over the **S**hort or **L**ong term (and **O**ngoing)
V = Vulnerability S = Strength

Top Priority Hazards (sea level rise, wildfire, flooding, tornado, ice, heat wave, hurricanes, etc.)

V or S	Vulnerabilities and Assets	Location	Ownership	Storm Surge	Inland Flooding	Wind	Sea Level Rise	Priority H - M - L	Time S/L term Ongoing
	Infrastructure								
V	Coastal Communities	Various	Private	Assessment needed that identify opportunities to improve resilience to various surge and sea level rise scenarios; Purchase property in floodplain; Increase amount of open space to limit ongoing risk.				H	S
V	School Facilities	Specific	Private	Assessment vulnerabilities to facilities under various scenarios; Identify ways to retain floods on recreational fields.				H	S
V	Wastewater Treatment Plant	Coastal	City	ID illegal connections; Determine vulnerabilities under various scenarios.			Install bioswales for inundation.	H	S-L
V/S	Coastal Zoning Policies	City-wide	City/Private	Assess strengths & weaknesses of existing zoning - various hazard scenarios.				M	Ongoing
S	Removal of Dam Structure	Specific	City/State/Private	Reduced extent of flood zone and enable restoration of connectivity upstream.					Ongoing
	Societal								
V	Inter-City & Regional Tabletop Exercise	City-wide	City	Need to conduct exercises to maximize readiness; Better regional planning/communication plan to discuss vulnerabilities, share ideas, resources, etc. Need to factor in the use of natural infrastructrue into emergency management procedures - evacuation, safe sites, etc.				H	S
V/S	Shelter capacity	Shelters	City	Coordinate with faith-based organizations to increase capacity to provide shelter during emergencies.				M	Ongoing
S	Disaster Recovery Task force (pre-deployment)	City-wide	City	Improve & enhance knowledge regarding recovery and building officials. Educate Task Force members on services and co-benefits of using natural infrastructrue in the rebuilding process.					Ongoing
S	Fewer people paying flood insurance	Waterways	City	Reduced number payering flood insurance (economic burden) by improving flood storage, storm surge protection, and open space conservation.					Ongoing
	Ecosystem								
V	Salt Marsh Restoration	Various	Private/State	Increase the restoration and enhancement of salt marsh habitat impacted by last storm to ensure adjoining properties are secured.				H	S
V/S	Beaches and Dunes	Various	State/Municipal	Assess risk reduction potential from existing and future wetland advancement zones. Restore/maintain beaches and dunes.				H	L
V/S	Riparian Corridors	Various	Muicipal/Private	Install green infrastructure projects such as bioswales and bank stabilization to increase viability of river systems and capacity to reduce flooding and erosion.				M	S
S	Advancement Zones	Various	State/Municipal	Secure upland advancement zones to accommodate future habitat that provide protective services and co-benefits.					Ongoing

Fig. 10.17 Risk Matrix used during Community Resilience Building Workshops to generate comprehensive and prioritized resilience action plans across and between sectors (infrastructure, societal, ecosystems) that incorporate natural infrastructure and eco-engineering. (Author's own graphic)

prioritization for alternative risk reduction approaches including natural infrastructure and hybrid solutions. With this community-based engagement platform in place there was a need to regionalize the dialogue to ensure broader action on resilience. In direct response to this need, the Regional Framework for Coastal Resilience in Southern Connecticut was launched in 2014 in the aftermath of Tropical Storm Irene and Sandy that hit the eastern seaboard of United States. Population centers within this Regional Framework including Greater New Haven and Bridgeport recognized a significant vulnerability to extreme weather events, sea-level rise, inland flooding, erosion, and coastal change. To counteract foreseeable risks, this project funded by the US Department of Interior, advocates "smart" investments in proactive risk assessment, community engagement, and conceptual design of on-the-ground projects to help minimize the consequences of synoptic scale storms and at the same time strengthen the resilience of existing and future ecosystems. The main objective of the project is to comprehensively assess, prioritize, and advance resilience opportunities to reduce risk to residents across ten coastal municipalities (30 % of the state's coastline) using natural infrastructure and hybrid solutions. This project incorporated the categorical and typological calibration steps for eco-engineering discussed above using 332 localized, risk reduction projects identified in the Regional Framework for Coastal Resilience. Furthermore, this project provides details on all 332 projects on a web-based tool platform (i.e., Coastal Resilience) via a "Regional Resilience Project" application.

This Regional Framework will include an integration of natural infrastructure concepts, principles, and priority projects into core hazard mitigation, comprehensive planning and capital expenditure efforts at the municipal and regional scale. This unique and critical initiative is being achieved through an active and growing partnership between regional planning organizations, municipalities, nongovernmental organizations, federal agencies, academic institutions, youth engagement organizations, businesses, and civic groups. To ensure receptivity for natural infrastructure, the highest priority projects across ten municipalities (20 of the 332 individual projects) were subjected to a conceptual design process driven by in-depth workshops and interviews. Conceptual design approaches used typologies of diverse, yet common circumstances faced by coastal communities from reactivating an urban industrial edge to restoring a public beach and dune complex to expanding the flood storage capacity of riparian corridor to living shorelines (Fig. 10.18). In all cases, this suite of projects and derived typologies integrate natural infrastructure in ways that increases the viability of the natural resource, provides improvements in public amenities (i.e., recreation, access, etc.), and reduce the risk from hazards to people over immediate and longer time horizons. This "triple-bottom line approach to resilience" using natural infrastructure is a hallmark of this project and a critical consideration for all other planning efforts. Ultimately, this type of approach will help accelerate the mainstreaming of natural infrastructure as a viable risk reducing option for communities across the globe.

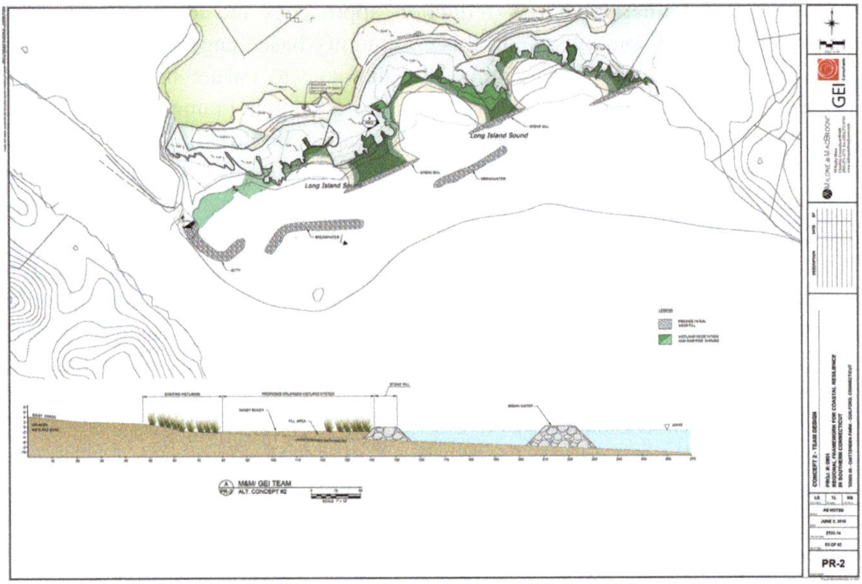

Fig. 10.18 Conceptual design of coastal eco-engineering approach using hybrid solution (i.e., living shoreline plus submerged breakwater structure). © GEI/Milone & MacBroom/TNC

Conclusion

To chart an alternative path for the globe and all societies present there is an immediate and longer term need to explore every opportunity to reduce risk and improve resilience. One clear opportunity lies in a broad, collaborative exploration of the science, policy, education, economics, and funding needed to mainstream the use of natural solutions via eco-engineering and design. There is clearly an unprecedented global interest and demand for advancing resilience to natural disasters and climatic change which universally recognizes the important contribution that ecosystems can and are making to reduce risk at global, national, and local scales. The overview, considerations, framework, supportive case studies, and recommendations presented herein will hopefully help move this imperative for humanity forward.

References

1. UNESCAP (2015) Overview of natural disasters and their impacts in Asia and the Pacific, 1970–2014. ESCAP Technical Paper—Information and Communications Technology and Disaster Risk Reduction Division. http://www.unescap.org/sites/default/files/Technical%20paper-Overview%20of%20natural%20hazards%20and%20their%20impacts_final.pdf. Accessed 12 June 2016

2. Kron W (2013) Coasts: the high-risk areas of the world. Nat Hazards 66(3):1363–1382
3. Reguero BG, Losada IJ, Diaz-Simal P, Mendez FJ, Beck MW (2015) Effects of climate change on exposure to coastal flooding in Latin America and the Caribbean. PLos ONE 10(7). http://journals.plos.org/plosone/article?id=10.1371/journal.pone.0133409. Accessed 25 June 2016
4 Woodruff JD, Irish JL, Camargo SJ (2013) Coastal flooding by tropical cyclones and sea-level rise. Nature 504:44–52
5. Hallegatte S, Green C, Nicholls RJ, Corfee-Morlot J (2013) Future flood losses in major coastal cities. Nat Clim Change 3:802–806
6. UNISDR (2011) Global assessment report on disaster risk reduction revealing risk redefining development. United Nations Office for Disaster Risk Reduction. http://www.preventionweb.net/english/hyogo/gar/2011/en/home/executive.html Accessed 5 July 2016
7. Bartlett S (2008) The implications of climate change for children in lower-income countries. Child Youth and Environ 18(1):71–98
8. Costello A (2009) Managing the health effects of climate change. Lancet 373:1693
9. Penrose A, Takaki M (2006) Children's rights in emergencies and disasters. Lancet 367 (9511):698–699
10. UNISDR (2015) Making development sustainable: the future of disaster risk management, global assessment report on disaster risk reduction. United Nations Office for Disaster Risk Reduction
11. International Council for Science (2015) International Social Science Council. Review of targets for the sustainable development goals: The science perspective. http://www.icsu.org/publications/reports-and-reviews/review-of-targets-for-the-sustainable-development-goals-the-science-perspective-2015/SDG-Report.pdf. Accessed 22 Jul 2016
12. Kelman I (2015) Climate change and the sendai framework for disaster risk reduction. Int J Disaster Risk Sci 6(2):117–127. http://link.springer.com/article/10.1007/s13753-015-0046-5. Accessed 19 July 2016
13. UN (2015) Paris Agreement. http://unfccc.int/files/essential_background/convention/application/pdf/english_paris_agreement.pdf. Accessed 22 July 2016
14. Renaud FG, Sudmeier-Rieux K, Estrella M (eds) (2013) The role of ecosystems in disaster risk reduction. United Nations University Press, Tokyo
15. Organization for Economic Cooperation and Development (2006) Infrastructure to 2030: telecom, land transport, water and electricity. http://www.oecd.org/futures/infrastructureto2030/infrastructureto2030telecomlandtransportwaterandelectricity.htm. Accessed 22 July 2016
16. Association Society of Civil Engineers (2013) Report Card for America's Infrastructure. http://www.infrastructurereportcard.org/. Accessed 22 Jul 2016
17. Jones HP, Hole DG, Zavaleta ES (2012) Harnessing nature to help people adapt to climate change. Nat Clim Change 2:504–509
18. Spalding MD, McIvor A, Beck MW, Koch E, Möller I, Reed D, Rubinoff P, Spencer T, Tolhurst T, Wamsley T, van Wesenbeeck B, Wolanski E, Woodroffe C (2014) Coastal ecosystems: a critical element of risk reduction. Conserv Lett 7:293–301
19. Temmerman S, Meire P, Bouma PJ, Herman PM, Ysebaert T, De Vriend HT (2013) Ecosystem-based coastal defence in the face of global change. Nature 5:504(7478):79–83
20. Arkema KK, Verutes GM, Wood SA, Clarke-Samuels C, Rosado S, Canto M et al (2015) Embedding ecosystem services in coastal planning leads to better outcomes for people and nature. PVAS 112(24):7390–7395
21. World Bank (2016) Managing coasts with natural solutions: guidelines for measuring and valuing coastal protection services of mangroves and coral reefs. Beck MW and Lange G-M (eds) Wealth Accounting and the Valuation of Ecosystem Services Partnership (WAVES). World Bank, Washington, DC
22. Beck MW, Brumbaugh RD, Airoldi L, Carranza A, Coen LD, Crawford C, Defeo O, Edgar GJ, Hancock B, Kay MC, Lenihan HS, Luckenbach MW, Toropova CL, Zhang G,

Guo X (2011) Oyster reefs at risk and recommendations for conservation, restoration, and management. Bioscience 61:107–116
23. Ferrario F, Beck MB, Storlazzi C, Micheli F, Shepard C, Airoldi L (2014) The effectiveness of coral reefs for coastal hazard risk reduction. Nat Commun 5(3794):1–9
24. Moller I, Kudella M, Rupprecht F, Spencer T, Paul M, van Wesenbeeck BK, Wolters G, Jenson K, Bouma TJ, Miranda-Lange M, Schimmels S (2014) Wave attenuation over coastal salt marshes under storm surge conditions. Nat Geosci Lett 7:727–731
25. Sheppard C, Crain C, Beck MW (2011) The protective role of coastal marshes: a systematic review and metaanalysis. PLoS ONE 6(11):272–283
26. CCRIF (2011) Annual Report 2010–2011. http://www.ccrif.org/sites/default/files/publications/CCRIFAnnualReport2010_2011.pdf. Accessed 23 July 2016
27. Warner JF, van Buuren A, Edelenbos J (eds) (2013) Making space for the river: governance experiences with multifunctional river flood management in the US and Europe. IWA Publishing, London
28. Millennium Ecosystem Assessment (2005) Ecosystems and human well-being: synthesis. Island Press, Washington, DC. http://www.millenniumassessment.org/documents/document.356.aspx.pdf. Accessed 21 July 2016
29. UNEP-WCMC (2011) Marine and coastal ecosystem services: Valuation methods and their application. UNEP-WCMC Biodiversity Series No. 33. http://www.unep.org/dewa/Portals/67/pdf/Marine_and_Coastal_Ecosystem.pdf. Accessed 21 July 2016
30. FAO (2016) The State of World Fisheries and Aquaculture 2016. Contributing to food security and nutrition for all. http://www.fao.org/3/a-i5555e.pdf. Accessed 20 July 2016
31. Costanza R, Pérez-Maqueo O, Martinez ML, Sutton P, Anderson SJ et al (2008) The value of coastal wetlands for hurricane protection. Ambio 37:241–248
32. Barbier EB (2007) Valuing ecosystems services as productive inputs. In: Beck T, Fuchs-Schundeln N, Gurkaynak R (eds) Economic Policy. Center for Economic Studies, University of Munich. (86)
33. Bateman IJ et al (2013) Bringing ecosystem services into economic decision-making: land use in the United Kingdom. Science 341(6141):45–50
34. Goldstein JH et al (2012) Integrating ecosystem-service tradeoffs into land-use decisions. Proc Natl Acad Sci USA 109(19):7565–7570
35. Guerry AD et al (2012) Modeling benefits from nature: using ecosystem services to inform coastal and marine spatial planning. Int J Bio Sci Ecosystem Serv Manage 8(1–2):107–121
36. Liquete C et al (2013) Current status and future prospects for the assessment of marine and coastal ecosystem services: a systematic review. PLoS ONE 8(7):67737
37. Tallis H, Polasky S (2009) Mapping and valuing ecosystem services as an approach for conservation and natural-resource management. Ann NY Acad Sci 1162:265–283
38. Samonte-Tan GPB et al (2007) Economic valuation of coastal and marine resources: Bohol Marine Triangle, Philippines. Coast Manag 35:319–338
39. Nishikawa S (2015) Incorporating science and technology for disaster risk reduction—The Japanese Experience. Planet@Risk 3(1):95–106
40. Inoue K (2007) Flood disaster in Japan. J Disaster Res 2:3–10
41. Nakagawa H, Kawata Y, Inoue K, Tanino T (1996) Flooding in Sandwip Island in the 1991 storm surge disasters. J Jpn Soc Nat Disaster Sci 15:151–174
42. Pingping L, Bin H, Kaoru T, Xiong Y, Nover D, Duan W, Fukushi K (2015) Historical assessment of Chinese and Japanese flood management policies and implications for managing future floods. Environ Sci Policy 48:265–277
43. Mitsch WJ (2004) Ecological engineering and ecosystem restoration. Wiley, NY
44. Narayan S, Beck MW, Reguero BG, Losada IJ, van Wesenbeeck B et al (2016) The effectiveness, costs and coastal protection benefits of natural and nature-based defenses. PLoS ONE 11(5). http://journals.plos.org/plosone/article?id=10.1371/journal.pone.0154735. Accessed 11 July 2016

45. Bueno R, Herzfeld C, Stanton EA, Ackerman F (2008) The Caribbean and climate change: the costs of inaction. Stockholm: Stockholm Environment Institute and the Global Development and Environment Institute, Tufts University
46. McIvor A, Spencer T, Moller I, Spalding M (2016) Coastal defense services provided by mangroves. In: Beck MW, Lange GM (eds) Managing coasts with natural solutions: Guidelines for measuring and valuing the coastal protection services of mangroves and coral reefs. World Bank, Washington, DC
47. Gaunnel G, Arkema K, Ruggiero K, Verutes P (2016) The power of three: coral reefs, seagrasses and mangroves protect coastal regions and increase their resilience. PLos ONE 11 (7)
48. Yang SL, Shi BW, Bouma TJ et al (2012) Estuaries and Coasts 35:169. http://link.springer.com/article/10.1007/s12237-011-9424-4
49. van Loon-Steensma JM, Hu Z, Slim PA (2016) Modelled impact of vegetation heterogeneity and salt-marsh zonation on wave damping. J Coast Res 32(2):241–252. http://www.jcronline.org/doi/abs/10.2112/JCOASTRES-D-15-00095.1. Accessed 22 July 2016
50. Hoover M, Whelchel AW (2015) Tidal marsh classification approaches and future marsh migration mapping methods for Long Island Sound, Connecticut, and New York. In: Tiner RW, Lang MW, Klemas VV (eds) Remote sensing of wetlands: applications and advances. CRC Press, Boca Raton
51. Horton R, Yohe G, Easterling W, et al (2014) Chapter 16: Northeast. Climate change impacts in the United States. In: Melillo JM, Richmond TC, Yohe GW (eds) The third national climate assessment. U.S. Global Change Research Program, pp 371–395
52. Ryan A, Whelchel AW (2015) A salt Marsh Advancement Zone Assessment of Connecticut. The Nature Conservancy, Coastal Resilience Program. Publication Series #1 Final (A-W), New Haven, Connecticut. http://coastalresilience.org/project-areas/connecticut-solutions/. Accessed 10 July 2016
53. Alongi DM (2009) Energitics of mangrove forests. Springer International Publishing, New York
54. Halide H, Brinkman R, Ridd P (2004) Designing bamboo wave attenuators for mangrove plantations. Indian J Geo-Marine Sci 33(3):220–225
55. Lopez JA (2009) The multiple lines of defense strategy to sustain Coastal Louisiana. J Coastal Res: Spec Issue 54:186–197
56. LACPR (2009) Louisiana Coastal Protection and Restoration Final Technical Report. U.S. Army Corps of Engineers, New Orleans District, Mississippi Valley Division. http://biotech.law.lsu.edu/la/coast/lacpr/FinalReport/03%20LACPR%20Final%20Technical%20Report.pdf. Accessed 25 June 2016
57. Shreve CM, Kelman I (2014) Does mitigation save? Reviewing cost-benefit analyses of disaster risk reduction. Int J Disaster Risk Reduction 10(A):213–235
58. Dedeurwaerdere A (1998) Cost-benefit analysis for natural disaster management. A case-study in the Philippines. Centre for Research on the Epidemiology of Disasters, Working Paper#143
59. IFRC (2009) Assessing quality and cost benefit: a Philippines case study
60. Johnston, RJ, Whelchel AW, Makriyannis C, Yao L (2015) Adapting to coastal storms and flooding: report on a 2014 Survey of Old Saybrook Residents. George Perkins Marsh Institute, Clark University and The Nature Conservancy, Connecticut Chapter. Worcester, MA. http://www.conservationgateway.org/ConservationPractices/Marine/crr/library/Documents/2014_Old_Saybrook_Report_FINAL_LoRes.pdf. Accessed 16 July 2016
61. Johnston RJ, Whelchel AW, Makriyannis C, Yao L (2015) Adapting to coastal storms and flooding: report on a 2014 survey of waterford residents. George Perkins Marsh Institute, Clark University and The Nature Conservancy, Connecticut Chapter. Worcester, MA. http://www.conservationgateway.org/ConservationPractices/Marine/crr/library/Documents/2014_Waterford_Report_FINAL_LoRes.pdf. Accessed 16 July 2016
62. Whelchel AW, Beck MW (2016) Decision tool and approaches to advance ecosystem-based disaster risk reduction and climate change adaptation in the twenty-first century. In: Renaud F,

Sudmeier-Rieux K, Estrella M, Nehren U (eds) Ecosystem-based disaster risk reduction and adaptation: liking science, policy and practice. Springer International Publishing, New York
63. Kroeker KJ, Reguero BG, Rittelmeyer P, Beck MB (2016) Ecosystem services and coastal engineering tools for coastal protection and risk reduction. In: Beck MW, Lange GM (eds) Managing coasts with natural solutions: Guidelines for measuring and valuing the coastal protection services of mangroves and coral reefs. World Bank, Washington, DC
64. Arkema KK, Guannel G, Verutes G, Wood SA, Guerry A, Ruckelshaus M, Kareiva P, Lacayo M, Silver JM (2013) Coastal habitats shield people and property from sea-level rise and storms. Nature Climate Change 3:913–918
65. Sharp R, Chaplin-Kramer R, Wood S et al (eds) (2014) InVEST user's guide: integrated valuation of environmental services and tradeoffs. Stanford University, The Natural Capital Project
66. Beck MW, Gilmer B, Whelchel AW, Ferdaña Z, Stone J, Raber G, Shepard C, Meliane I, Newkirk S (2013) Using interactive decision support to integrate coast hazard mitigation and ecosystem services in Long Island Sound, New York and Connecticut USA. In: Renaud F, Sudmeier-Rieux K, Estrella M (eds) Linkages between ecosystems, livelihoods and disaster risk reduction. UNU Press, Bonn, PEDRR, pp 140–163
67. Ferdaña Z, Newkirk S, Whelchel AW, Gilmer B, Beck MW (2010) Building interactive decision support to meet management objectives for coastal conservation and hazard mitigation on Long Island, New York, USA In: Andrade Pérez A, Herrera Fernandez B, Cazzolla Gatti R (eds) Building resilience to climate change, Gland, IUCN, Switzerland, pp 72–87
68. Airoldi L, Beck MW (2007) Loss, status and trends for coastal marine habitats of Europe. Oceanogr Mar Biol Annu Rev 45:345–405

Author Biography

Adam Whelchel—Director of Science, The Nature Conservancy Dr. Adam Whelchel's twenty-six year career in leadership positions has catalyzed partnerships with strategic direction while with government agencies, private firms, academic institutions, and non-profits in the United States, China, Southeast Asia, Europe, Caribbean, and Africa. Adam Whelchel is an author, facilitator, and dynamic communicator called upon to assist diverse groups from around the world with team building, strategic planning, and community resilience. Adam is also a senior scientist and practitioner immersed in the fields of ecological and urban resilience. Recent accomplishments include serving as a key facilitator for the Puerto Rico Climate Adaptation Project, lead author for the U.S. National Climate Assessment, and architect of Community Resilience Building.

Chapter 11
Protect the Ocean, Protect Ourselves

Sylvia Earle

From space, it is obvious: The world is blue. Shrouded by water as a vapor in clouds, crowned with water as diamond-bright polar ice, and wrapped with water as a liquid that embraces all land masses—water sets Earth apart from other places in a beautiful yet inhospitable universe.

> Suddenly, from behind the rim of the Moon, in long, slow-motion moments of immense majesty, there emerges a sparkling blue and white jewel, a light delicate sky-blue sphere laced with slowly swirling veils of white, rising gradually like a small pearl in a thick sea of black mystery. It takes more than a moment to fully realize this is Earth… home.

S. Earle (✉)
Mission Blue and SEAlliance, P.O. Box 6882, Napa, CA 94581, USA
e-mail: info@mission-blue.org

M. Johnson and A. Bayley (eds.), *Coastal Change, Ocean Conservation and Resilient Communities*, DOI 10.1007/978-3-319-41914-5_11

It is water, filled with life that makes our lives on this blue planet possible, represents biodiversity, feeds us, and sustains vitally important systems and economies. With every drop of water you drink, every breath you take, you're connected to the sea—no matter where on Earth you live. Most of the oxygen in the atmosphere is generated by creatures living in the sea. Every fish fertilizes the water in a way that generates the plankton that ultimately leads back into the food chain, but also yields oxygen and grabs carbon; it is a part of what makes the ocean function and what makes the planet function. The ocean drives climate and weather, stabilizes temperature, and shapes Earth's chemistry. Water from the sea forms clouds that return to the land and the sea as rain, sleet and snow, and provides home for about 97 % of life on Earth. The ocean is the cornerstone of our planet's life support system, and the cornerstone of the ocean's life support system is life in the ocean.

The poet Auden said, "Thousands have lived without love; none without water." If you think the ocean isn't important, imagine Earth without it. Mars comes to mind. No blue, no green.

Most of the ocean—covering 71 % of Earth's surface—is cold and dark, but it is *alive*. Looking at a spot on a map, if you only know what is at the surface it is difficult to tell what is really going on any more than you would know about New York City if you just looked at the tops of the buildings. It's not just water and rocks. It's a living system; the water itself is full of life.

My first glimpse beneath the waves was as a fearless three-year-old, tumbled by a rogue wave of green water. Cresting white, the Atlantic Ocean engulfed me, sweeping me upside down, pulling me deeper, holding me, then shifting, surging me back to where my toes touched sand, and strong arms lifted me to safety. A moment later, I jumped back in, exhilarated. The ocean has held my attention ever since.

When I was 12, my family moved from New Jersey to Florida and the Gulf of Mexico became my backyard. I spent hours just searching through the giant piles of seaweed that washed ashore, picking up tiny crabs and releasing them back into the sea. The Gulf Coast was a vast wilderness. It once seemed that—as with the ocean as a whole—the Gulf was so big, so vast, so resilient, that nothing we could do could harm it. The benefits we believed would always be there, no matter how large the trawls, how long the nets, how numerous the hooks for catching ocean wildlife—or how many, how long or how deep the pipelines, drilling operations, seismic surveys, or production rigs.

I experienced my first breath underwater as a young scientist in 1953, and I marveled at the clarity of the ocean and the wealth and diversity of life during trips to the Florida Keys. Pink conchs plowed trails though seagrass meadows, and schools of colorful fish crowded the branches of elkhorn and staghorn coral. Long, bristly antennae marked the presence of lobsters under ledges and crevices, and elegantly striped and irrepressibly curious Nassau grouper followed me on most dives and likely would have continued onto the beach but for the limitations of fins and gills.

Mesmerized by ocean pioneer Jacques Cousteau's writings and films of the 1950s, I, too, wanted to explore the clear blue Mediterranean Sea, glittering with hundreds of kinds of fish, forested with angular corals, giant sponges, and rainbow colored plankton. From my family's home in Dunedin, Florida, my "Mediterranean" was the Gulf of Mexico, similarly populated with several kinds of sea turtles, grouper, snapper, coral reefs, sea grass meadows, and shallow mangrove-fringed coastal areas in places sloping off into dark waters thousands of feet deep.

Most of what is known about the ocean is in its upper 1000 feet or so. Despite the years of research by hundreds of scientists and institutions, knowledge about the nature of our ocean is still primitive, partly because the methods used for exploring the ocean are still primitive. During the Gulf oil disaster, Larry McKinney observed that we know more about the face of the moon than the bottom of the Gulf of Mexico, and are better equipped to live and work in space than we are to explore the ocean on this planet. Only about 10 % of the ocean has been mapped with the same degree of resolution that we have for the land, or the moon, or Mars, or Jupiter.

If you dive in a submarine, even in broad daylight, you see more shades of blue than there are even *names* for blue—from pale to ultradark, ultraviolet, indigo, then deep black. When I had the chance to go walking 381 meters deep on the seafloor in the JIM diving suit in Hawaii in 1979, I had the chance to explore the deep sea autonomously with only a communication cable connecting me to the world above. I imagined that, being down that deep, everything would be completely dark—just black—but it wasn't. Since I was diving at midday, there was enough sunlight filtering down through the clear ocean water that it looked like the deepest indigo. Bioluminescent creatures flashed with their blue fire—fish with blinking lights, flashing blue octopuses—some of the same creatures that had enchanted me as a child from reading the works of William Beebe. I saw a shark with a luminous green eye and a field of bamboo coral—single, whisker-like spirals that grew nearly six feet tall like tall white bedsprings with big bands of black. So many wonderful creatures, and so many yet to be discovered.

The ocean may seem big and deep, but it is not infinite. We humans have this idea that the Earth—all of it: the ocean, the atmosphere—is so vast and so resilient it doesn't matter what we do to it. That may have been true 10,000 years ago, but in the last 100, especially in the last 50, we've drawn down the assets, the air, the water, the wildlife that make our lives possible. Now we know: Coral reefs, kelp forests, coastal marshes, numerous kinds of fish, and other ocean wildlife have declined sharply owing to pressures we have applied. Dead zones have appeared. Oxygen-producing plankton is declining. Ninety percent of all large wild fish (and many small ones as well) have disappeared from the ocean—the result of our giant appetites and devastatingly efficient industrial fishing. Seawater is warming and acidifying from our carbon dioxide emissions. Globally, about half of the coral reefs that existed when I was a child are gone or are in a state of serious decline.

Six decades after taking my first breath underwater, I note the difference. The waters of the reefs where I made some of my earliest dives are not nearly as clear as they are in my recollections. The great forests of branching corals are largely gone. The pink conchs and Nassau grouper are mostly memories; the remaining few are protected in U.S. waters because of their rarity. With care, there is a chance that these and many other species may recover, but some losses are irrecoverable. I missed meeting, for instance, one of Florida's most charismatic animals, the

Caribbean monk seal, a playful St. Bernard-sized creature that once lolled on beaches throughout the region, sometimes ranging as far north as Galveston, Texas. The last one was sighted in 1952. The species is now officially listed as extinct.

How long before the fish are all gone if we keep eating them this way, and how long before all of the corals bleach and reefs erode, destroying habitat for countless species?

The impact that humans have had on our planet since I was a child is greater than during all preceding human history put together. For one thing, there are more of us. When I arrived, there were 2 billion people and now there are over seven billion; and the planet has not gotten any larger. The pressure that we are putting on the land, the air, the water, and the wildlife that keep us alive has been stressed significantly. And as a witness to the changes, I feel compelled to share the view—especially what has happened to the sea.

Whatever else we achieve, the ultimate success will be to dispel ignorance about the sea. Of all the ocean's problems, what we don't know poses the greatest threat. Knowledge is the key to making a difference. With knowledge, you can be powerful. My goal is to push that frontier of ignorance further and deeper, and to return to the surface brimming with knowledge.

Perspective has arrived just in time. Now we know what no one could understand until recent times: The ocean is Earth's blue engine, driving climate and weather, regulating and stabilizing planetary temperature, governing global chemistry, including the basic cycles of water, oxygen, carbon, nitrogen, phosphorous, and more. And there is good news. Ten percent of the big fish still remain. Half the coral reefs are still in pretty good shape, a jeweled belt around the middle of the planet. There's still time, but not a lot, to turn things around. But business-as-usual means that in 50 years, there may be no coral reefs and no commercial fishing, because the fish will simply be gone.

Now we know—the ocean is not too big to fail. There is only so much that we can take out of the ocean before it will be empty of fish and other sea life, and there is also a limit to how much of our greenhouse gas emissions, excess toxic chemicals and trash we can throw into the ocean before this vital living system will no longer be able to function.

The ocean is in trouble and that means we are in trouble, too. Native Hawaiian navigator and president of the Polynesian Voyaging Society, Nainoa Thompson, says it this way: "There's only one island. It's blue, and it's called Earth." We are all sea creatures. Protecting the ocean really protects ourselves. Take away the ocean and we don't have a planet that works. Take away life in the ocean and we don't have a planet that works. No ocean, no us. No blue, no green.

With knowledge of what is driving the changes that are altering the nature of the ocean, we have the power to reverse the unfavorable trends. Some important actions are underway. The establishment of "marine protected areas" (or MPAs)—designated zones of ecological importance in which activities such as fishing and mining are strictly prohibited much like national parks on land—has become a very popular conservation method that has power to protect the health of the ocean.

My dream is for everyone to use all means at their disposal to spark a movement to create a global network of marine protected areas—"Hope Spots"—large enough to save and restore the ocean, the blue heart of the planet. I founded Mission Blue to ignite support for this goal. By designating Hope Spots around the world, we are highlighting places in the ocean in need of special protection with the goal of safeguarding at least 20 % of the ocean by 2020, and 30 % by 2030. Mission Blue is committed to protecting the ocean, Earth's blue heart—and in so doing, provides hope for an enduring future for humankind.

Hope Spots are areas in the ocean that have been identified for protection owing to their magnified importance in protecting, restoring and maintaining the health of the ocean. Mission Blue has established nearly 60 Hope Spots to date, and more are being added as new areas and opportunities are identified. Some are large, some small; some are pristine, whereas others are greatly stressed. But with care, all provide hope for steady improvement in the current state of ocean health and resilience. According to analyses by the International Union for Conservation of Nature, about 12 % of Earth's terrestrial area is designated for parks, reserves, and wildlife management zones, but for the sea, less than 4 % has equivalent care. Less than 2 %of the ocean is fully protected. For successful recovery of ocean species and systems, much more is needed.

Slowly but surely, several nations are showing leadership in increasing ocean protection. The tone was set in 2006 by two presidents: George W. Bush, who designated major areas in the northwestern Hawaiian Islands and the western Pacific, and Anote Tong, leader of the Pacific island Republic of Kiribati, who declared protection that year and in 2008 for 158,000 square miles of ocean surrounding the nation's 33 atolls and islands. Another island nation, the United Kingdom, followed in 2010 with what at the time was the world's largest fully protected marine reserve: 225,810 square miles around the Chagos Archipelago in the Indian Ocean. In 2012, Australia announced that it would create a network of marine reserves covering 888,035 square miles of sea and bringing the total area of Australia's protected ocean to 1.2 million square miles. In 2014, President Obama expanded the areas established by President Bush to create one of the largest protected areas on Earth. And 2015 was a banner year for marine protection, with over one million square miles of highly protected ocean announced that year alone, protecting more of the planet than any other time in our history. The governments of Palau, Great Britain (with overseas territory surrounding Pitcairn and Ascension Islands), and Chile (with overseas territory surrounding Easter Island) announced massive marine protected areas.

World leaders are also gathering at meetings like the U.S. Department of State's Our Ocean Conference and at the United Nations to negotiate protections for the High Seas, which cover half of our planet yet have barely been explored—especially at depth. What especially sets the high seas apart from all other marine areas is the nearly complete lack of protection for any of this natural heritage. We owe it to ourselves and to future generations to change that. We can prevent the extinction of countless species and ecosystems that are only just being discovered, let alone understood. In conservation, action often comes after destruction has occurred. For the high seas, the United Nations is in a unique position to act before irreparable damage is done.

People are waking up to the reality that NOW is the time to act. Researchers and advocates are working hard to ban practices like shark finning and bottom trawling, and they are making progress. There is still a *long* way to go, but people are waking up to the fact that the ocean is not an inexhaustible resource. We must make it uncool to eat tuna, or swordfish, or grouper, or other ocean wildlife. We need to give them a break. We used to worry about *man-eating* sharks. Now we worry about man, *eating sharks*. *Sharks* are not *eating* us; we're *eating* them. In the Stone Age, we had no choices. We had lots of wildlife and few of us. Now that we have over seven billion people with wildlife under great stress, we need to protect every fish alive because they are much more valuable as part of the natural world that keeps us alive than they are cooked on a plate.

Now we know: Everyone can make a difference just by changing what they consume. Less demand will mean less support for commercial fisheries and the fossil fuel industry. It isn't too late to shift from the swift, sharp decline of ocean systems in recent decades to an era of steady recovery. There is time, and there is a growing awareness, which is the best way to counter indifference.

We need a global movement of people using their unique talents and skills to create innovative solutions to climate change and policies that fully protect vast networks of ocean areas, both near-shore and in the high seas. We need people from all backgrounds and professions to raise awareness and inspire empathy about threats like climate change, ocean acidification, and overfishing among their communities. Researchers need to speak out about their findings; activists need to spread the word about them; policy makers need to hear from voters and corporations that saving the ocean is a priority, and they need to work with scientists to act on those demands in effective ways. The only difference that has been made ever in the world, for good or for not so good, always starts with just one person. But it will take a coalition of researchers, indigenous communities, students, engineers, explorers, artists, teachers, and advocates to use their unique skills and new technologies to appeal to our global society and change our relationship to the ocean for good.

Fifty years ago, we could not see limits to what we could put into the ocean, or what we could take out. Fifty years into the future, it will be too late to do what is possible right now. We are in a "*sweet spot*" in time. We need to convey a sense of urgency because the world is changing quickly. Never again will there be a better time to take actions that can insure an enduring place for ourselves within the living systems that sustain us. We are at an unprecedented, pivotal point in history when the decisions we make in the next 10 years will determine the direction of the next 10,000. We must take care of the ocean as if our lives depend on it, because they do.

Author Biography

Sylvia Earle—Oceanographer is a National Geographic Society Explorer-in-Residence and the former Chief Scientist of the National Oceanographic and Atmospheric Administration (NOAA). She is the founder and chairman of the Sylvia Earle Alliance / Mission Blue, founder of Deep Ocean Exploration and Research, Inc., Chair of the Advisory Council of the Harte Research Institute, inspiration for the Ocean in Google Earth, and subject of the Emmy Award winning Netflix documentary, Mission Blue. Dr. Earle's current focus is on sparking support for a global network of marine protected areas—"Hope Spots" in the ocean to safeguard the living systems that provide the underpinnings of global processes, from maintaining biodiversity and yielding basic life support services to providing stability and resiliency in response to accelerating climate change. Her research concerns marine ecosystems with special reference to exploration, conservation, and the development and use of new technologies for access and effective operations in the deep sea and other remote environments. She has a B.S. degree from Florida State University, M.S. and PhD. from Duke University, and 29 honorary degrees.

www.nationalgeographic.com/explorers/bios/sylvia-earle/.

Chapter 12
Concluding Statement: Saving Coastal Ecosystems and Our Life Support System While We Save Our Waterfront Cities

Marcha Johnson

The challenge of coastal changes especially sea level rise is not just, *how do we build resilient city structures so that storms and flooding do less damage?* But also *how can we continue to live on the coastal edge so that "the cornerstone of earth's life support system"* [1]—*the oceans—are healthy*?

Some initial steps

- Relocate a majority of the pavements, buildings especially homes, critical infrastructure and most of the man-made features now in floodplains, on beaches and on barrier islands which are incompatible with salt water and intermittent flooding to higher land. Releasing the coastal zone from most development will not only take the expensive and vulnerable farther from harm's way, but also reestablish an environment with multiple benefits to us, and to the natural environment.
- Employ non-structural approaches such as zoning, building codes, and evacuation planning in planning resiliency.
- Conserve and rebuild vegetative buffers along the water's edge that absorb and reduce waves, such as mangrove forest and marshes.
- Manage fishing practices so that fish populations can rebound.
- Allow beaches, salt marshes and coastlines to migrate, both inland and parallel to the shore. This could greatly reduce the incentive to "nourish beaches" in ways that harm both life on the beach and at the dredged sand's source.
- Maintain habitat for the high diversity of species which use edge zones of overlapping water, air, and earth.
- Monitor and manage city shores so that ecosystem services such as water filtration, oxygen production, and maintenance of healthy nutrient levels are priorities.

M. Johnson (✉)
ASLA, Brooklyn, NY, USA
e-mail: marcha.johnson@gmail.com

M. Johnson and A. Bayley (eds.), *Coastal Change, Ocean Conservation and Resilient Communities*, DOI 10.1007/978-3-319-41914-5_12

Shall we continue to consider humans as the controllers of nature and the natural world as "ours" to manipulate for our own benefit? There are still many examples of a continuation of this attitude, in particular in the response to sea level rise and coastal storms. We are hopeful that alternative approaches will evolve and prevail, following many of the examples showcased in this book, in which we learn to reverse poor choices, and figure out how to have thriving coastal cities where marine life and marshes thrive, too.

Reference

1. Sylvia Earle (2016) http://www.ngllife.com/sylvia-earle-1. Accessed 30 Apr 2016

Index

M. Johnson and A. Bayley (eds.), *Coastal Change, Ocean Conservation and Resilient Communities*, DOI 10.1007/978-3-319-41914-5

MIX
Papier aus verantwortungsvollen Quellen
Paper from responsible sources
FSC® C105338

FSC
www.fsc.org

If you have any concerns about our products,
you can contact us on
ProductSafety@springernature.com

In case Publisher is established outside the EU,
the EU authorized representative is:
Springer Nature Customer Service Center GmbH
Europaplatz 3, 69115 Heidelberg, Germany

Printed by Libri Plureos GmbH
in Hamburg, Germany